Praise for *Compost Scienc*

Do you want to know anything about compost or composting? And I do mean *anything*! Robert Pavlis's book *Compost Science for Gardeners* covers it all, including of course, leading you through the nitty gritty, hands-on making of compost, various kinds of bins, and best uses of compost. Also, some specialized composts, such as bokashi and vermicompost, and more.

— Lee Reich, PhD., author, *Growing Figs in Cold Climates*
and *The Ever Curious Gardener*

Myth-buster Robert Pavlis has done it again, writing an essential resource on the science (and art) of composting. This will be a great addition to your gardening bookshelf.

— Rebecca Martin, technical editor, *Mother Earth News* magazine

Robert Pavlis does what he does best—using down to earth science to teach, this time to make great compost. This should be your go-to manual for great soil!

— Jeff Lowenfels, author, *DIY Autoflowering Cannabis*,
and the Teaming book series on organic growing

Robert Pavlis is always thorough, accurate and his work very readable. This volume on compost will sit on our shelf of "most frequently visited" books where we can refer to it often. Composting should not be complicated. Reading *Compost Science for Gardeners* takes the hocus pocus out of the process and helps gardeners make the most of their composting efforts. Thank you Robert! This book is long overdue.

— Mark and Ben Cullen, Cullen's Foods

Compost Science for Gardeners

compost
science
for gardeners

Simple Methods
for Nutrient Rich Soil

ROBERT PAVLIS

new society
PUBLISHERS

Cover design by Diane McIntosh.
Cover image © iStock

Printed in Canada. First printing November 2022.

Inquiries regarding requests to reprint all or part of *Compost Science for Gardeners* should be addressed to New Society Publishers at the address below. To order directly from the publishers, please call 250-247-9737 or order online at www.newsociety.com.

Any other inquiries can be directed by mail to:

New Society Publishers
P.O. Box 189, Gabriola Island, BC V0R 1X0, Canada
(250) 247-9737

LIBRARY AND ARCHIVES CANADA CATALOGUING IN PUBLICATION

Title: Compost science for gardeners :
simple methods for nutrient rich soil / Robert Pavlis.

Names: Pavlis, Robert, author.

Description: Includes bibliographical references and index.

Identifiers: Canadiana (print) 20220421056 | Canadiana (ebook) 20220421196 |
ISBN 9780865719767 (softcover) | ISBN 9781550927702 (PDF) |
ISBN 9781771423663 (EPUB)

Subjects: LCSH: Compost—Handbooks, manuals, etc. |
LCSH: Garden soils. | LCSH: Gardening—Handbooks, manuals, etc.

Classification: LCC S661 .P38 2023 | DDC 631.8/75—dc23

New Society Publishers' mission is to publish books that contribute in fundamental ways to building an ecologically sustainable and just society, and to do so with the least possible impact on the environment, in a manner that models this vision.

Contents

1

Introduction

What is the secret to great flowers and more vegetables?

You can buy great starter plants or high-quality seeds, plant them in the right amount of sun or shade, and water correctly, but all of that has a limited effect on plant growth. The secret to great plants is the soil. Get the soil right and you can grow anything that is hardy in your location.

The obvious next question is, how do you get great soil? The answer to that question is a bit more complicated, but a key ingredient is organic matter. Adding organic matter to soil increases microbial activity, releases plant nutrients, and improves soil structure.

That nice crumbly black gold that gardeners talk about is the result of higher levels of organic matter in the soil.

Nature adds organic matter to soil all the time. Fall leaves blanket the ground, and by the following summer they have been magically incorporated into soil. Animals run through the area adding some fresh manure, and many insects die due to short life spans, adding even more organic matter. The grasses in fields set down deep roots which are constantly dying off and regenerating, all the while adding organic matter.

We see all of these processes taking place, but few of us think about the way in which organic matter is cycled around. It starts as CO_2 in the air, which is absorbed by plants and combined with sunlight to form sugars and other carbohydrates. These high carbon compounds form the basis of all organic matter.

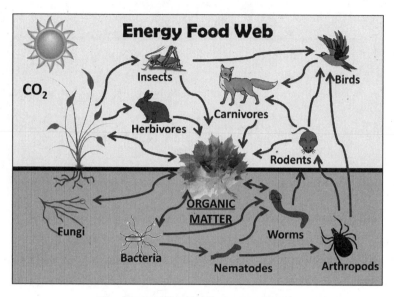

The Energy Food Web moves carbon
from the air into the soil.

When plant material falls to the ground, it becomes a carbon food source for microbes. They soon convert that plant into invisible organic matter, all the while moving it deeper into the soil.

If the plant is eaten by animals, insects, or worms, it is converted into fecal matter which is nothing more than partially digested organic matter. Some of the carbon in the food is digested and absorbed by the animal. All non-plant organisms, including microbes, animals, insects, birds, and even humans, are essentially digested organic matter that originated from plants. In the end, they are returned to the soil.

All of the processes that I have just described also happen in your garden, and you can improve on them or inhibit them. It's your choice.

Some gardeners get in the way of these natural processes. They keep their soil in pristine condition, not allowing any old vegetation from being incorporated into soil naturally. They spray for pests and reduce the number of insects that live and die in the garden. They grow vegetables, harvest the produce, and take all of the old plant material to the curb for disposal. Fall leaves are raked and given to the city.

This makes for a very neat garden, but over time the soil has less and less organic matter, making it unhealthy.

It does not have to be that way, and many gardeners take a different approach. They do the opposite. Plant matter is left where it falls. Most insects are treasured and encouraged to use the property. Some gardeners even go so far as to collect bags of leaves from neighbors and bring them back to their garden. A big part of their gardening focus is to add more organic matter than they had when they started.

There is also a third group, the impatient gardeners. They don't want to wait for nature to incorporate yard waste and kitchen scraps into soil. They want to speed up this process. Composting is the way to do that.

Composting is nothing more than helping nature speed up the decomposition process. It takes fallen leaves, dead insects, kitchen scraps, and manure and accelerates the process of turning that material into nutrients for microbes and plants.

All of the composting methods described in this book make use of natural processes, but gardeners manipulate them so they are more efficient. The end product is essentially the same as the organic matter produced by nature.

What Is Compost?

Merriam-Webster defines compost as a mixture that consists largely of decayed organic matter and is used for fertilizing and conditioning land. This definition is not correct since the organic matter is not fully decayed.

Wikipedia defines compost as a mixture of ingredients used to fertilize and improve soil. Compost is used to fertilize, and it does improve soil, but so do other things, like manure and wood chips, but neither of these are compost.

I define compost as partially decayed organic matter that feeds plants, feeds the soil biology, and improves soil.

Why Should Everyone Compost

If all residents composted, the annual volume of waste pickup would be dramatically reduced. The EPA said, "Food scraps and yard waste together currently make up more than 30 percent of what we throw away, and these materials could be composted instead."

Sending this organic material to landfill sites not only fills them up faster, it ends up producing a potent greenhouse gas called methane which is 25 times worse than CO_2 for global warming. Not to mention all the pollution caused by trucks running all over town collecting material.

Gardeners who compost need to buy fewer fertilizers, which is a benefit to the environment *and* their pocket book. Even non-gardeners who only have a lawn would save the expense of fertilizer.

Composting can be quite simple, and in the ideal world everyone would be required to compost or at least give their organic material to a neighbor to compost. Municipalities should not have to deal with this material.

Myths About Composting

There are a number of composting myths that might keep you from trying it. Let's dispel them right now.

Myth #1: It's too difficult.
Fact: It is as simple or complex as you want to make it. Follow my cut and drop method and it is actually easier than what you are doing right now. At the other extreme, you can become very serious and make perfect hot compost. You decide on how much effort you want to put into the process.

Myth #2: Compost piles smell.
Fact: They only smell if you do something wrong. When composting is done right, it has a very pleasant odor and smells like a walk in

the woods after a rain or, in the case of bokashi, it smells a bit like pickles.

Myth #3: It costs a lot of money.
Fact: You can spend money on it, but you can also compost for free. I made compost bins using some free skids and a bit of wire.

Myth #4: Composting attracts rodents.
Fact: Animals do live in the garden so you might have some rodents, but if you keep meat and dairy products out of the compost, rodents are rarely a problem. Bears might be a bigger problem if you compost outside.

Myth #5: It takes a lot of space.
Fact: It does take some space depending on the system you use. My cut and drop method takes no space at all. Vermicomposting can be done indoors. Some systems use very little space.

Myth #6: You need an outdoor garden.
Fact: I'll present several composting methods that work indoors.

Definitions

I like to start with definitions because they frame the discussion and ensure that we are all on the same page. In the UK and probably in other parts of the world, the word "compost" is used to describe potting soil. This is not what we are talking about here. In this book, the word "compost" refers to the end product of a composting process, which consists of taking plant material and other compostable material, piling it up, and letting it decompose into a black friable material.

Aerobic vs Anaerobic

Composting can take place in two different conditions: aerobic and anaerobic. An aerobic environment is one where oxygen is plentiful. These systems are usually open to air which is what provides the oxygen.

Anaerobic conditions are ones where there is very little oxygen. These can be created in one of two ways. Some systems are filled with water which pushes the air out and therefore keeps the oxygen levels low. Another way to create such a condition is to use a closed vessel. Decomposition uses up O_2 and produces CO_2. In a closed vessel, this results in air that has higher levels of CO_2 and very little oxygen, creating an anaerobic condition.

Why is this important? Bacteria tend to live in one or other of these conditions. Some are aerobic and some are anaerobic, but just to keep things interesting, some can live in both conditions and are called facultative anaerobes.

By controlling the amount of available oxygen, gardeners can control the type of bacteria working in the compost, and that in turn controls the type of decomposition that takes place.

What Is Organic?

Organic is a word that is used, far too frequently, to mean several different things, so it's not always clear what it means. This leads to all kinds of misunderstandings.

The term "organic" has become synonymous with the word "natural," which leads to the misconception that anything organic is good for us, our garden, and the planet. The term is used extensively to describe products, in the hope that people will buy them. In the same vein, "organic" has also come to represent non-synthetic chemicals. The reality is that many natural organic chemicals are more toxic than synthetic ones. Most drugs are synthetic and, for the most part, are safe and beneficial. Some natural organic chemicals, such as ricin, which is found in the caster bean, is one of the most toxic compounds on Earth. Organic does not mean safe.

Organic is also used to refer to agricultural produce that is grown "organically." This does not mean they are produced without pesticides or chemicals. It just means that when chemicals are used, they fall under a strict set of guidelines developed by certified organic organizations. If you follow their rules, you are organic, even if some of the approved chemicals are synthetic or toxic. The rules become paramount and safety is secondary.

To a chemist the word "organic" means something completely different. An organic chemical is any chemical that contains carbon and is not a salt. All sugars, carbohydrates, proteins, and most pesticides contain carbon and are organic, even if they are man-made.

This book will use the chemist's definition of organic and the term "certified organic" to refer to organic agriculture.

I'll use the term "organic matter" in a very general sense to refer to any dead flora, fauna, or microbe. This could be recent dead material such as wood chips and manure, or a highly decomposed form such as compost or humus. Composting starts with organic matter, and at the end of the process, you still have organic matter. Composting changes its characteristics, but finished compost is still organic matter.

2
The Role of Compost in Soil

Compost is critical for developing soil structure and feeding plants. This chapter will help you understand it better and explore the impact it has on soil.

The word "compost" is used in a very general way in this book. It refers to plant and animal material that has undergone extensive decomposition so that the original material is no longer recognizable. It is normally dark in color, quite friable, and smells good and earthy.

Composting is the process of making compost. The more traditional way of doing this is to pile up the organic matter and let it rot. Over time the natural microbial biology converts the starting organic matter into compost.

As you will see in this book, there are many ways to make compost, and there are many types of starting material. But that does not mean there are many types of compost.

We don't look like plants, but on a molecular basis, and even at the cellular level, we are not that different. In fact 50% of our DNA is the same as a plant's DNA. That number is a bit misleading since a lot of that is not active, but the point is valid. Chemically, plants and animals have a lot of similarities.

When cells are decomposed, they produce proteins, carbohydrates, vitamins, hormones, and fats (oils). Further decomposition degrades these large molecules even more into simple compounds like amino acids, simple sugars, and nutrients. At this point in the decomposition process, there is virtually no difference between you, me, or a banana.

The majority of all living organisms are made up of carbon, hydrogen, and oxygen. When this composts fully, it forms humus.

The reason all compost is basically the same is that once all of this decomposition has taken place, we are left with the simple molecular building blocks that are used in all forms of life. Compost can have more or less water in it, it can have higher or lower nutrient levels, but the similarities are far greater than the differences.

Gardeners commonly ask which compost is best. They go to the store and see piles of composted cow manure and composted horse manure, so they naturally wonder which one is better? The answer is, they are the same. It's all compost.

The best compost is the one you can get in larger quantities and at a low price. This almost always means it is transported the shortest distance, which is good for your pocket book as well as the environment.

The very best compost is the one you make yourself. It is mostly free, reduces material going to landfill, and you know exactly how it was made.

Benefits of Composting

Compost has many benefits for the gardener. This section is an overview of these benefits, which are discussed in more detail throughout the book.

Compost as a Mulch

Used as mulch, compost provides most of the benefits of other types of mulches. It keeps the soil cool, which plant roots love, and it reduces evaporation, which keeps the soil moisture more constant and reduces watering frequency.

It may or may not help with weeds. It will keep seeds at the soil surface dark, preventing them from germinating. Unfortunately, any new seeds that land on top of the compost will find a perfect place to grow. Some compost can also contain viable seeds, in which case they will easily germinate in the mulch.

Compost as a Fertilizer

Compost contains a good supply of plant nutrients including all of the micronutrients. It also releases these nutrients over time, acting like a slow-release fertilizer. It is perfect in areas where you want a low but steady long-term feed, like most ornamental beds.

Cation exchange capacity (CEC) is a measure of the soil's ability to hold nutrients. A higher CEC means that the soil can hold more, which is good for soil fertility. Soil with a low CEC holds fewer nutrients and results in poor plant growth.

Compost has a high CEC which means nutrients stick to it and are more slowly leached away. This is especially important for sandy soil that is unable to hold on to nutrients.

Compost as a Soil Builder

Compost is the best option for improving soil. If you have ever gone into the woods and felt the black crumbly soil you find there, you'll know what good soil is—what we gardeners call black gold. That soil is the result of natural compost being added over hundreds or even thousands of years.

Compost helps build aggregation and improves the structure of both sandy soil and clay soil. Nothing improves soil better than compost.

Compost Retains Water

Compost holds a lot of water, keeping it near the soil surface where plant roots can get to it. It helps maintain an even moisture level thereby reducing watering needs.

A 5% increase in organic material quadruples the soil's ability to hold water. Compost holds water equal to 200% of its own dry weight.

Compost Removes Toxins

It's like a sponge for heavy metals like lead and cadmium. Compost grabs hold of them as they float by in the soil solution and holds on to them so that plants are exposed to lower levels.

Compost can also hold onto other toxins, like pesticides.

Compost Buffers pH

Both acidic and alkaline soils are neutralized, bringing the pH level closer to the optimum range for plants (6 to 7).

Compost Increases Microbes in Soil

The secret to healthy soil are the microbes. More microbes translate into better soil, which results in better plants. Compost not only provides food for microbes but it also gives them a place to hide.

Compost Helps the Environment

Composting uses organic waste material that would normally end up in landfill. This reduces the amount of material going to landfill while at the same time reducing greenhouse gases like methane.

The Microorganism Myth

Microorganisms (microbes) are responsible for improving the structure of soil. They also decompose organic matter to release nutrients that plants can use. They are critical for the health of soil and plants.

One way of determining soil health is to measure the number of microbes. The more microbes in soil, the healthier the soil.

Knowing this fact, people automatically assume that adding more microbes to soil will improve it, and most information online or in books supports this idea. Manufacturers even sell bottled microbes to make it easy to "improve soil." The sad fact is that this is all a big myth.

Adding microbes to soil does not increase the number or diversity of microbes living in soil.

You might be sitting there shaking your head thinking that can't be true. If you add more, you obviously have more, which is actually true for a short while. However, in a matter of hours, the number of microbes will be back to the level you had before you added them.

Think of soil as being a football stadium. It has 50,000 seats and every seat is full. Another 5,000 sport fans arrive looking for seats, but they are all full, so the latecomers can't stay.

If you have crappy soil, your stadium may only hold 10,000 seats, but they are always full. Good soil has 80,000 seats, and they are always full too.

Microbes multiply very quickly, with some species doubling in number every 20 minutes. The number of microbes in soil at any given time depends on moisture, temperature, air, and available food. Of these, food is the most important parameter controlling populations. If available food increases, the population explodes. As food levels drop, microbes die off.

The key point is that at any given point in time, the population is at a maximum level for the current food, air, temperature, and moisture levels. The available seats are always full.

Improving any of the four parameters will result in a higher number of microbes, but just adding more microbes will have no effect.

What happens when compost is added to soil? It contains microbes and microbe food, the organic matter. Just like soil, the compost also has all of its seats filled with microbes. Adding this to soil does increase the total number of microbes but only to the extent that the organic matter can support them.

What Is Soil Health?

The term is used a lot but what does it really mean? Depending on your interest, it can mean different things. A climate scientist might define "healthy soil" as one where the amount of sequestered carbon is increasing. A farmer might define it as soil that produces a good yield. A microbiologist may be measuring microbe population and diversity.

Gardeners look at plant health. If a plant is growing well, flowering a lot, and has no diseases, the soil must be healthy or at least healthy enough to grow good plants. Some plants grow well in nutritious soil, while others grow much better in sandy, lean soil. The definition of soil health depends very much on the type of plant you are growing.

I am not going to provide a specific definition, but for the purpose of this book, healthy soil is one that grows a wide range of plants, has good aggregation, has a good amount of organic matter, and supports a high number of microbes. Admittedly, that is a squishy definition, but it is good enough for our purpose.

Aggregation is a measure of how well the base ingredients of soil—sand, silt, clay, and organic matter—are clumped together to form larger soil particles. You can see and feel good aggregation by handling soil from an undisturbed forest or meadow. It's easy to dig, consists of larger crumbly pieces of soil that allow a lot of air to get in, and the spaces between the particles are large enough for root growth.

What Is Soil?

As you can see from the above list of benefits, compost plays a huge role in soil development. In order to understand this in more detail, it is important to have a better understanding of soil. This section will introduce a number of soil topics, but due to limited space, it can't cover all of the things gardeners should know. If you want to know more about soil, have a look at my other book, *Soil Science for Gardeners*.

Soil consists of mineral components and organic matter. The mineral components include sand, silt, and clay. You can think of sand as being small stones and silt as extremely small stones. Sand is much larger than silt, but both particles have the same physical and chemical properties. Neither one holds water or nutrients very well.

Clay is made up of very small particles, much smaller than silt. These particles have electrical charges on them that hold onto nutrients. The spaces between particles are also very small, and they easily fill with water. That water is very slow to drain away. That is why clay soil tends to stay wet while sandy soil dries quickly.

Most gardeners know that organic matter is important to soil, but they are surprised to learn that good soil only contains about 5% organic matter. In fact, most soil at newer homes, and agricultural soils, have much less than this. The value is even lower in sandy soil, which is usually closer to 1%.

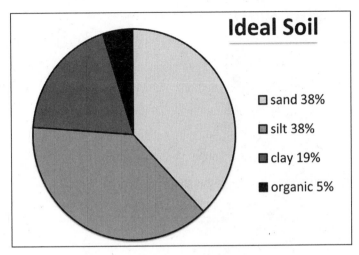

Components of ideal soil.

The diagram above shows the amount of sand, silt, clay, and organic matter in ideal soil. It is important that you realize nobody has ideal soil, and it is not your job to create ideal soil. You are stuck with the sand, silt, and clay in your garden, and you have to make the best of it. You can, however, increase your organic matter, and this book shows you how to do that. Even a change of less than 1% can make a big impact on the health of your soil.

Air and Water

Air and water are critical for proper plant growth, and they make up about 50% of soil. The actual amount depends on several factors such as soil texture, the amount of organic material, and the degree of compaction, but ideal soil contains about 25% air and 25% water.

Immediately after a heavy rain, much of the air has been forced out and replaced with water. Gravity, evaporation, and plants will then reduce the level of water, which is replaced with air. Perfectly dry soil will have no water and 50% air. Such dry soil is rare and is mostly found in laboratories. Soil normally holds some water even if plants are no longer able to get any of it.

Evaporation is the process where liquid water turns into water vapor and escapes into the air. This happens mostly at the surface of

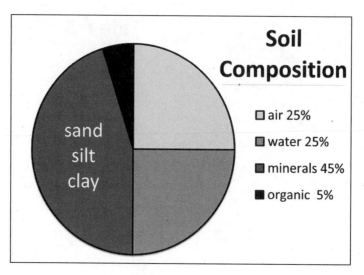

Soil composition.

the soil, and that is why the top layer of soil can be quite dry while, a few inches down, the soil is still quite wet.

As evaporation takes place, more water will be drawn to the surface by capillary action. Over time this process slowly dries out the soil. Mulch slows down evaporation, keeping the soil moist for a longer period of time.

Gravity is also at work, and it pulls water deeper into the soil. Eventually water is pulled down far enough to enter reservoirs deep in the ground, or depending on topography, it might flow into a river or lake.

Plant roots are constantly absorbing water and transferring it to their leaves, where much of it evaporates through leaf openings called stomata. Water is also used in chemical processes like photosynthesis. Removal of water by plants can be significant. A large tree can remove up to 100 gallons (400 liters) of water a day, and discharge it into the air as water vapor.

Organic matter plays a critical role in this water-air cycle by slowing down both evaporation and water loss due to gravity. It does this by holding on to the water molecules more tightly than soil, thereby slowing its movement both up and down.

Organic matter is full of tiny pores that hold lots of air and water. When it is added to clay soil, it increases the amount of air in the soil allowing it to dry faster. In sandy soil it holds water longer, keeping the soil from drying out.

Compost is a perfect form of organic matter, and even small amounts can improve the water-air cycle to benefit plants and microbes.

Aggregation and Soil Structure

So far we have looked at soil on a microscopic level, but soil is much more than that. If you pick up some good soil, you will notice that it does not look like a bunch of sand, silt, and clay. What you see is larger crumbly pieces which are dark in color. The space between particles is quite large. The soil is fluffy with lots of air pockets of all different sizes.

What you are looking at is the macro structure of good soil. The smaller pieces of sand, silt, and clay have been mixed with organic matter to form larger structures called aggregates. My woodland soil has aggregated particles of one-quarter inch (six millimetres) in size and many can be a full inch in diameter.

Aggregation is not well understood by gardeners, but it is a very critical part of soil health. When soil has it, you have good soil that will grow lots of plants. When aggregation is lacking, the soil performs poorly. Creating good soil is all about improving aggregation.

The key to aggregation is a special binding agent that consists of many different kinds of chemicals produced by living organisms. Think of them as life juices. Plants, bacteria, fungi, earthworms, and small insects all excrete juices, and some of these chemicals work great as a glue. Fungi and actinomycetes make even larger aggregates by using their mycelium to knit smaller particles together.

Organic matter is the food for all of these organisms. A higher level of organic matter in soil translates into a higher microbe population which results in more glue and better aggregation.

Why are aggregates so important? Clay and silt are very small particles, and the spaces between them are too small for roots to penetrate. As soil aggregates into larger particles, the spaces between particles gets larger, which makes it easier for plant roots to grow. The small pores inside each aggregate are also perfect for bacteria to hide from larger organisms that want to eat them.

Aggregation is a continuous process that is either improving or getting worse. The natural binding agents slowly decompose and need to be continually replaced by microbe activity. An annual addition of compost ensures that aggregation is always improving.

Plant Nutrients

Gardening is all about growing plants, and as plants grow, they absorb nutrients from air and soil to form larger molecules and cellular structures. Composting is the opposite. Composting takes the complete plant, or animal, and decomposes them back to basic nutrients. This is called the nutrient cycle.

Plant nutrients can be broken down into two main categories: mineral nutrients (originating from minerals) and non-mineral nutrients.

The non-mineral nutrients make up 96% of a plant and consist of oxygen, hydrogen, and carbon. The plant absorbs CO_2 from

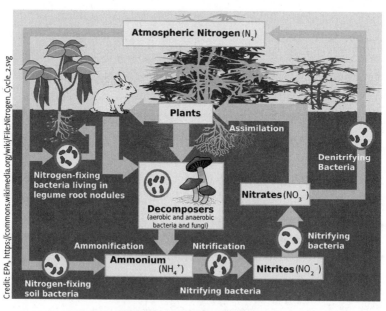

Credit: EPA, https://commons.wikimedia.org/wiki/File:Nitrogen_Cycle_2.svg

Nutrient cycle for nitrogen.

the air which provides most of the carbon and some of the oxygen. They also absorb water and oxygen through the roots. The water is transported to leaves where it is broken down into oxygen and hydrogen. A properly watered plant has no problem getting all of the non-minerals it needs.

The mineral nutrients only account for 4% of a plant's weight and include things like nitrogen, phosphorus, and potassium. Although they make up a small part of a plant, the mineral nutrients are vital to its growth.

Both soil particles and organic matter contain mineral nutrients, but plants can't use them until they are released in a plant-available form called ions. Once in the form of ions, they dissolve in the soil solution (the water in the soil) and are available to plants.

Ions

You probably recognize the names of many micronutrients such as iron, zinc, potassium, calcium, and magnesium, but what you may not know is that these are all metals. For example, pure calcium is a dull silvery-gray color and looks a lot like iron.

We talk about plants using iron, calcium, and magnesium, but the truth is that plants can't use any of these. Putting an iron nail in soil does nothing to feed plants, so it can't solve nutrient deficiencies, although that myth is common in social media. Plants are only able to use these metals once they are converted into ions.

When calcium is exposed to air, it reacts with the oxygen to form a type of rust. This chemical reaction produces a white powder called calcium oxide which contains one molecule of calcium and one molecule of oxygen (CaO). When calcium oxide dissolves in water, something special happens. The calcium and oxygen separate into electrically charged particles called ions. The calcium ion has a positive charge (a cation), and the oxygen ion has a negative charge (an anion).

Once the calcium is in the form of an ion, plants are able to absorb it through the roots and use it inside the plant. We talk about plants using calcium, but what they really use are calcium ions. This

Ionic Forms of Plant Nutrients

Nutrient name	Chemical symbol	Ion form	Ion name
Carbon	C	none	
Hydrogen	H	H^+	
Oxygen	O	O^{-2}	
Nitrogen	N	NO_3^-, NH_4^+	Nitrate, ammonium
Phosphorus	P	HPO_4^{-2}, $H_2PO_4^-$	Phosphate
Potassium	K	K^+	
Calcium	Ca	Ca^{+2}	
Magnesium	Mg	Mg^{+2}	
Sulfur	S	SO_4^{-2}	Sulfate

may seem like unimportant semantics, but it is critical for understanding how nutrients behave in soil, and how they become available to plants.

All of the mineral nutrients used by plants form some type of ion. Some are a bit more complex than the above calcium example, but the principles are exactly the same. This table shows a list of the macronutrients and their ion forms.

In the early stages of composting, you find big pieces of organic matter, a whole leaf or an apple core. These all contain nutrients, but they are tied up in the form of large molecules and complete cells. None of these nutrients are available to plants.

As decomposition takes place, cells and large molecules are broken down into smaller and smaller pieces until ions are released. *It is only then that plants can use these nutrients.*

What Is Salt?

The general public uses the term "salt" to mean table salt, which is sodium chloride. Chemists, soil scientists, and this book use the term to refer to any compound that is made up of ions.

Compost Myth: Salt Kills Soil Microbes

Many believe that the salts in synthetic fertilizer harm soil life, but that is not true. The salts in fertilizer dissolve in water, forming ions. These ions are exactly the same as the ions released from organic material like manure or compost. They are essential for the growth of microbes and plants.

Any chemical, no matter how useful, can become toxic at high levels. Even too much water will kill you. Provided fertilizers (i.e., salts) are used in appropriate amounts, they do not harm soil life.

Sodium chloride is one of many different types of salt. In water, it breaks up into sodium ions (Na^+) and chloride ions (Cl^-).

The calcium oxide discussed above is also a salt. Compounds such as ammonium nitrate and potassium phosphate, found in synthetic fertilizer, are also salts. Urea fertilizer is an organic molecule made up of carbon, hydrogen, oxygen, and nitrogen, and since it does not form ions in water, it is not a salt.

Table salt or road salt releases sodium ions in water. All life, including plants, need some sodium, but as the amount of sodium in soil increases, it can quickly become toxic to plants and microbes. It is best to keep sodium out of the garden.

Movement of Nutrients in Soil

Sand and silt particles have almost no electrical charge on their surface so ions don't stick to them very well. When nutrient ions come in contact with these particles, they just keep moving along with the water. This is the reason why rain easily washes nutrient ions out of sand and silt into the subsoil layers, and explains why such soils have low natural fertility.

Clay and organic matter have charges on them that act like little magnets. These magnets attract both anions and cations and hold them tightly, preventing water from washing them away.

Compost Myth: Organic Nutrients Are Better

This is a very common myth that is promoted by the organic movement. They believe that nutrients from organic sources are much better for plants than nutrients from synthetic fertilizer. This concept is completely wrong.

Synthetic fertilizer usually consists of simple salt compounds. Good examples are ammonium nitrate, calcium carbonate, and potassium phosphate. When these compounds dissolve in water, they separate into ions, namely ammonium, nitrate, calcium, carbonate, potassium, and phosphate. Plant roots can absorb all of these.

When an organic source, like manure or compost, is added to soil, it slowly decomposes into the same ions found in synthetic fertilizer. The nitrate ion from an organic source is exactly the same as a nitrate ion from a synthetic fertilizer. Neither labs, nor plants, nor microbes can tell the difference between the two sources, once the ion has been released into water.

Once you understand this, it becomes clear that both sources result in exactly the same nutrients. Nutrient ions can originate from an organic source, but they can't be any more organic than the ones from fertilizer. They are all inorganic.

When rain flows through clay soil or soil that contains a lot of organic matter, it does dislodge some nutrients and moves them deeper in the soil, but the effect is minor. Most nutrients remain stuck in place. The net effect is that nutrients move much more slowly in these soils than in sandy soil.

Soil that contains more organic matter holds mineral nutrients better, making these soils more nutritious.

Tilling

The organic level in soil is constantly changing. Adding compost to soil will increase the level and, as you have seen, that has a positive effect on soil health. Gardeners can also do things to decrease the organic level, and one of the most controversial is tilling.

Tilling has historically been a standard practice. Each spring the vegetable garden is tilled to remove weeds and get the soil nice and fluffy, ready for planting seeds. Unfortunately this practice has some downsides. It destroys aggregation, which in turn makes surface crusting worse. It also brings weed seed from deeper in the soil up to the surface, creating more weed problems.

For a number of years, it was also believed that tilling reduces the level of organic matter in soil. Tilling adds more air to soil. With access to more air, microbe numbers start to grow and they break down the organic matter faster, and this has been confirmed through testing. However, more recently scientists have looked at deeper soil profiles and found the story is more complex.

Tilling does reduce the organic matter in the top 6 inches (15 cm) of soil, but it increases the level below the 6-inch mark. The total amount of organic matter in the top foot (30 cm) of soil is not changed with tilling. It could be that tilling moves more organic matter to lower levels, or it might be due to plants being able to grow deeper roots in tilled soil.

The addition of air to soil during tilling is also not as significant as first thought. Air is added for a short period after the tilling event, but over a period of months or years, this effect is insignificant compared to other processes.

Tilling is not as bad as originally thought, or as claimed by some fringe groups, but it is a practice that most gardeners can stop. Use mulch to keep weeds down, and make smaller beds so you don't have to walk on the planting surface. Combining these two techniques eliminates the need for tilling in established beds.

Mulching

Mulch is any material that covers the ground between plants. It will retain moisture, suppress weeds, keep the soil cool, and some feel it makes the garden look better. It can be either organic or inorganic, but organic material has the added bonus of improving the soil as it decomposes. The advantage of inorganic material is that it does not need to be replaced.

Effect of Fertilizer and Mulch on Soil Properties

	Fertilizer	Compost	Wood chips
Density	Lower +	Lower ++	Lower ++
Moisture	Same	Up +	Up ++
Organic matter	Same	Up ++	Up +
Respiration	Up +	Up ++	Up ++
pH	Same	Up ++	Up +
Nitrogen	Same	Up ++	Up +
Phosphorus	Up +	Up +++	Up +
Potassium	Same	Up +++	Up ++

Credit: GardenMyths.com based on data by Bryant C. Scharenbroch and Gary W. Watson

Compost Myth: There Is No Such Thing as Too Much Compost

Compost is organic and slowly provides nutrients to the garden. Since it is good for the garden, many people feel that more is better, but that is not true.

The problem lies with the nutrient ratios in compost compared to the nutrients used by plants. Plants use nitrogen, phosphorus, and potassium (NPK) in a ratio of about 3-1-2. They use three times as much nitrogen as phosphorus. Compost generally has an NPK ratio of 1-1-1, namely, equal amounts of each nutrient.

If you supply a plant with the correct amount of phosphorus using only compost, it will be lacking in nitrogen. If you use more so that the plant has enough nitrogen, you will be supplying way too much phosphorus, and this is what people tend to do. After a few years, their soil becomes toxic due to a high level of phosphorus.

Small amounts of compost (one inch a year) are good for the garden and will avoid toxic phosphorus levels.

As organic mulch decomposes, it has a significant effect on the soil. A five-year study[1] compared fertilizer, compost, and wood chip mulch, by measuring density (i.e., compaction), moisture, organic matter, respiration (microbe activity), pH, nitrogen, phosphorus, and potassium. It found that wood chip mulch increased the organic matter in soil, decreased compaction, increased microbial activity, and increased nutrient levels. These changes lead to better aggregation and better plant growth.

Mulch also reduces crusting, increases rain infiltration, and reduces runoff. Raindrops can no longer hit the surface of the soil, eliminating the crusting problem. The mulch also disperses raindrops, allowing it to more slowly percolate down to the surface of the soil, where it is easily absorbed.

Compost is very good mulch, one that provides a lot of nutrients and is very effective at improving the quality of soil. It is generally applied as a thin layer of about one inch and, as such, is not great for reducing weed growth. However, it is better to apply compost as mulch than to dig it into the soil.

The Science of Composting

Composting is all about creating an ideal environment for microbes. Supply them with the right temperature, nutrients, moisture, and oxygen, and they will do all the work for you.

This section takes a more detailed look at what happens during the composting process, and how various parameters affect the process. It is critical to understand this so that you can make informed decisions in selecting and managing the right composting process.

Composting starts with the raw materials. These are normally organic materials that you would easily recognize...an apple core, a banana peel, a dead tomato plant, grass clippings, some newspaper. You can recognize each of these input materials because they have a very visible structure.

If you looked at these materials on a cellular level, you would see a lot of detailed structures. The stem of the tomato plant has xylem and phloem structures, which are surrounded by epidermis (outer skin). Everything is made up of cells, which have their own structure.

Now let's look at this more closely on a molecular level. Everything is made out of complex molecules. Big proteins, containing thousands of atoms. DNA strands that are huge. Complex starches and oils.

The difference between a live tomato plant and a dead one is almost zero. There has been virtually no decomposition in that dead tomato plant, and under a microscope, it looks just like a live plant.

There are other ingredients as well: things like manure. It's not as recognizable because an animal has chewed it up and started the decomposition process in its stomach. You might not be able to see a plant stem, but on a molecular basis, most of the large molecules and cellular structures are still present.

Now let's jump ahead to the end of the composting process. All of the familiar structures of the plant are gone. The cells have all been broken apart. A lot of the large molecules have been decomposed. Proteins are now small amino acids, and some of those have even been converted to free nitrate molecules, which plants can use. Large starch molecules are now simple sugars. Plants don't use those sugars, but microbes do, and when they use the sugars, they end up releasing nutrients that plants can use.

In short, composting converts all of the large molecules in the input ingredients into small molecules that can be used by plants and microbes. Now, let's have a closer look at how we get from large molecules to small ones.

The Reality of Composting

Gardeners like to make composting much more complicated than it needs to be. If you leave plant material on the ground, it will decompose. If you pile it up, it will also decompose. If you put it in a fancy rotating drum or a four-foot-square bin, and get the greens and browns in exactly the right ratios, it will also decompose. The difference between these various methods is one of speed and temperature.

Composting happens when microbes, mostly bacteria and fungi, decompose organic matter. When the pile of material is large enough, and the carbon to nitrogen ratio is ideal, the process creates a lot of heat, and you end up with hot composting, which is fast and kills weed seed and pathogens.

In smaller piles, or when the C/N ratio is too high, the process is slow and called cold composting.

The microbe populations in compost are very dynamic. As the food source breaks down, different microbes get involved and others start dying off. The pH and temperature also change, and as they change, microbe populations change.

The reality is that organic matter will decompose and produce compost. Your actions can speed it up or slow it down, but eventually, it all composts, even if you do nothing.

The Hot Composting Process

All composting methods are similar, but to keep things simple, I will only deal with traditional hot composting in this section.

Composting is done by organisms, and each of these has a preferred temperature range where they grow best. Psychrophilic organisms prefer to stay below 50°F (10°C). The mesophilic prefer 50°F to 105°F (10°C to 40°C), and the thermophilic organisms like temperatures over 105°F (40°C) but below 150°F (66°C).

Microorganisms thrive and multiply in ideal temperatures. As temperatures become too low or too high, organisms slow down their activity and in extreme cases may die.

Hot composting goes through three distinct phases: the initiation phase, the thermophilic phase, and the maturation phase.

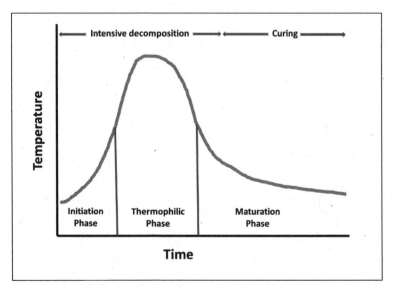

The hot composting process.

The Initiation Phase

Things start with a pile of organic matter. We'll talk about specific ingredients shortly but for now think of the pile as a mixture of all kinds of material.

One thing that surprises gardeners is that this material is naturally covered with billions and billions of microorganisms, including thousands of different species. Many are bacteria, but some are fungal spores or actinomycetes. These organisms are hungry and have already started digesting the organic material, but the pile is a more hospitable environment so they speed up the process.

If the air temperature is low, psychrophilic microorganisms including both bacteria and fungi start digesting material. All of these organisms breathe in oxygen and give off CO_2 as they metabolize (eat) the organic matter. They also give off a lot of heat in the process, and the pile starts to warm up, activating the mesophilic organisms which will dominate the pile until temperatures reach about 105°F (40°C). These organisms are the most efficient decomposers.

This initiation phase will last about 3 days in a well-built compost pile.

The Thermophilic Phase

The warmer temperatures are perfect for new types of microbes called thermophilic. This group includes a wide range of bacteria, fungi, and actinomycetes, and these organisms are responsible for most of the composting. A lot of the hard to break down cellulose is broken apart in this phase.

Temperatures approaching 150°F (66°C) will kill off many organisms, including pathogens. It will also kill weed seeds. Human skin can sustain first-degree burns at 118°F (48°C) and second-degree burns at 131°F (55°C), so be careful sticking your hand in a hot compost bin.

Many organisms form spores when the temperature gets too hot. These spores allow them to survive until conditions are more suitable for them.

Temperatures higher than 150°F (66°C) are rare in backyard

Compost Myth: Hot Compost Piles Can Spontaneously Combust

Hot compost produces a lot of heat and that causes steam to rise from the pile. Some equate this steam to smoke and think the pile will spontaneously combust, but that won't happen in a normal-sized pile. The material is too wet and the temperature is too low to ignite the material.

There have been cases of commercial compost piles catching fire, but they usually create much larger piles than homeowners.

Spontaneous combustion can occur if the right conditions are met.

- Temperatures reach 300°F (150°C).
- The pile is too large. For this reason, National Fire Protection Association (NFPA) regulation 230 suggests piles should not exceed 60 feet in height, 300 feet in width, or 500 feet in length.
- Moisture levels drop, creating dry material.
- Oxygen levels drop, leading to the production of methane gas, which has a low flash point.

It would be hard for a backyard pile to meet these conditions.

composters, but larger commercial piles can get warmer. Most operators prevent this by turning the pile since higher temperatures will also kill off the beneficial organisms.

As the process continues, food starts running low and microbe activity slows down. This causes the temperature of the pile to drop.

The thermophilic phase usually lasts a week or two.

The Maturation Phase

As the temperature starts to drop, mesophilic organisms once again thrive and they take over the pile. Overall activity is now much less than in the previous phase, but organic matter is further decomposed and some of it is converted to humus.

The maturation phase is the longest phase and can last several months.

Understanding Microbes

The microbes in a compost pile are very diverse and involve thousands of species. Each species has its own preferred living conditions and food source. It is important to understand that these are not "composting microbes" in the sense that they are only found in compost. These microbes are found all over your garden, on the plants, in water, and in the soil. You do not need special microbes to do composting.

The key players can be grouped into three categories: bacteria, fungi, and actinomycetes.

Bacteria

Bacteria are very small; 500,000 take up no more room than the period at the end of this sentence. A single teaspoon of soil contains as many as 30,000 species, and they make up 80% of the organisms in a compost pile. The sheer number of bacteria is astounding. A gram of fertile soil, about the weight of a paper clip, contains up to a billion bacteria. It is hard to get your head around such a big number but consider this: there are as many bacteria in three tablespoons of fertile soil as people on Earth.

They are single-celled and have a variety of shapes including spheres, rods, and spirals. When it is time to reproduce, they split into two cells, each being a copy of the original. In perfect conditions, this can happen every 20 minutes, producing huge amounts of bacterium in a short time. In a lab situation, a single bacterium can grow to five billion in 12 hours.

We tend to think of bacteria as being one type of organism, but the diversity is huge. There are species that live in virtually every type of environment on Earth, and eat almost everything, even spilt oil and jet fuel. Some like it hot, some like it cold, some wet, some dry. They are found everywhere, but each environment hosts a different community of species.

Most bacteria are heterotrophs, which are organisms that get their energy source from carbon. A few are autotrophs and get their energy from soil minerals.

Some bacteria use oxygen in the same way we do, and this group is called aerobic bacteria. Another group of bacteria prefer to live in environments that have a very low level of oxygen; these are anaerobic bacteria. There is even a group that can live in both environments. An example of this latter group is *E. coli*, a bacterium that lives in our anaerobic intestine, as well as in soil (which is usually aerobic).

Hot composting is an aerobic process and will select for those bacteria that live in such conditions. Other forms of composting that I'll explore later in this book, such as bokashi and eco-enzyme, are anaerobic.

Bacteria don't have mouth parts so they don't really eat, but they do need to get nutrients. They have a cell wall that keeps important chemicals in and keeps unwanted chemicals out. Bacteria use something called active transport to selectively move molecules through the cell wall. Think of it like the border around your country. There are controls in place that keep unwanted people out, but let others in. A bacterium wall works the same way. It lets nutrients in and keeps harmful molecules out.

This still leaves bacteria with a problem. Assume they are sitting next to some fresh yummy plant material. All of the nutrients in this material are tied up in large molecules, which are too large to transport across the cell wall. Instead of just waiting for something to happen, bacteria take aggressive action. They excrete a variety of enzymes through their cell wall. These enzymes decompose the organic matter, breaking the large molecules into smaller and smaller molecules until the nutrient ions are released. The bacterium then absorbs these nutrients.

Once inside the bacterium, the nutrients are immobilized into new large molecules.

The process seems very inefficient, but it must work reasonably well because bacteria are responsible for most of the decomposition in a compost pile and they produce most of the heat.

All of this activity—making enzymes, transporting molecules through cell walls, and dividing—requires a lot of energy. That energy comes from carbon containing molecules, such as sugar, that the bacterium must also absorb from its environment. This food source is as important as nutrients.

Organic matter contains a lot of molecules containing carbon, and these are released in the decomposition process. Almost 50% of a plant is made up of cellulose, and bacterial enzymes can break this down into sugars that provide a good energy source.

Lignin is a major component in wood, and it too contains a lot of carbon. Although bacteria can't decompose it, fungi can.

Fungi

The fungi are as important as bacteria. They are decomposers that carry out functions that bacteria are unable to do.

Fungi are a strange group of organisms. They have some plant-like characteristics, such as the hyphae, that function in a manner similar to roots and grow out into organic matter looking for nutrients. But fungi have no chlorophyll, so they can't make their own food. They depend on carbon compounds for their energy source, just like bacteria and animals.

Beer, wine, and bread are the result of the single-celled yeast fungi, and some commercial products such as probiotics contain fungi. The mold on old bread and cheese is also evidence of fungi. When you eat mushrooms, you are eating the fruiting bodies of multi-celled fungi.

Fungi start life as a spore, a special kind of seed that is normally produced by the fruiting body. The fine powder given off by an old puffball consists of spores. If you take a mushroom cap and let it sit for a day, you will find powdery spores under it.

Spores are very small and easily float through air. The breath you just took sucked in a bunch of them. They travel great distances on wind currents, which means that fungal spores are everywhere. They exist universally in all climates and can even be found in the Antarctic.

When these spores land in a suitable environment, they sprout and start growing hair-like hyphae. Hyphae are made up of long chains of cells that form a type of tube. Liquid, nutrients, and other compounds flow easily from one end of the tube to the other.

The tip of the hyphae is continually growing and branching to form new arms of hyphae that are all connected together. They can grow as fast as 40 micrometers per minute, about the thickness of a human hair. These hyphae tend to mass together to form larger clumps which are then called mycelium.

A single gram of soil, the weight of a paper clip, can contain 330 to 3,300 feet (100 to 1,000 meters) of hyphae. Fungi are less numerous than bacteria, but since each one is much larger, the total mass of fungi in soil is about the same as the mass of bacteria in soil, and together they account for most of the microbe population.

Over 100,000 species have been identified, of which 70,000 can be found in soil. It is estimated that the total number of species is around 1.5 million.

Fungi are heterotrophs, which means they cannot make their own food. They need to find a carbon source, and they do this in a way that is very similar to bacteria. They excrete various enzymes and other compounds that decompose the large organic molecules that surround them. Once converted to small nutrients and sugar molecules, these can be absorbed and transported along the hyphae, for several feet, to a point where they are needed.

Unlike bacteria, some fungi can digest the really tough organic matter like lignin and residues that are too dry, acidic, or low in nitrogen for bacterial decomposition. The tip of hyphae can produce enzymes that let it penetrate hard surfaces like plant leaves and stems to get at the more digestible material inside. Most of the absorption of food takes place near the tip of the hyphae.

A unique quality of fungi is their ability to grow hyphae above the soil line in order to penetrate leaves and other plant refuse lying on the surface of the soil. This is one way that fungi get into a compost pile or even a pile of leaves on the ground.

Fungi prefer temperatures of 70°F to 75°F (21°C to 24°C). They can't survive temperatures above 140°F (60°C).

Actinomycetes

Actinomycetes are commonly called mold bacteria and thread bacteria. They look and grow more like fungi, but biologically they are similar to bacteria. They grow hyphae-like threads that consume resistant organic matter (cellulose and lignin), and they are tolerant of dry soil, alkaline soil, and high temperature conditions. They are also responsible for the earthy smell of compost.

They produce chemicals that stop the growth of other microbes, such as streptomycin and actinomycin, which are commercially available antibiotics.

Actinomycetes tend to be found in decaying organic matter and can be visually seen as a gray-white web that looks like mold. Their affinity for higher temperatures and ability to decompose tough organic matter such as woody stems, bark, and even newspaper makes them an important element in hot composting, but they also play a role in cold composting and the maturation process.

The Role of Macroorganisms

Microbes get all the press, but macroorgansims also play a major role in composting. These include things like mites, centipedes, sow bugs, snails, beetles, ants, flies, and earthworms. These are physical decomposers that grind, tear, and chew material into smaller pieces. This provides better access to the food for the microorganisms.

These macroorganisms work well in a cold or cool compost pile, but most cannot survive the thermophilic phase. They will leave the pile as things warm up and return when the pile cools down.

Macroorganisms are also important for moving bacteria around. Their bodies are coated with them, and their intestines, like our own, are full of them. Think of macroorgansims as bacterial Uber drivers.

Although gardeners don't like some of these organisms, such as flies and slugs, they are an important part of composting and should be left alone. Ants will rarely build their nest in a pile unless it gets too dry for composting. The larvae of flies become food for beetles. It's all part of nature.

What Role Do Worms Play?

Gardeners love earthworms in their garden, and they know that worms eat organic matter and improve soil health. Worms are even used as the main decomposer in vermicomposting, and they are often found in great numbers in finished compost. So it is natural to think that adding worms to a compost pile would be a good thing, but it's not.

Earthworms grow best at low temperatures, and they are killed at 130°F (55°C). Clearly they won't last a long time in a hot compost pile. Earthworms will move into the pile on their own when it suits them.

The reason they are found in finished compost is that they tend to gather there at the end of the composting process to eat their favorite food: bacteria. Earthworms don't actually want to eat organic matter. They eat it to get at the bacteria on the organic matter, and there are a lot in finished compost.

Heat Kills Pathogens

Microorganisms produce antibiotics, and we have used this function to produce drugs to kill a number of human pathogens. This fact has led people to believe that the microbes in compost can produce enough antibiotics to kill pathogens, but this is not true. Pathogens are killed by heat and time.

At 115°F (46°C) pathogens are killed off in a week, while at 144°F (62°C) they are killed off in an hour. Any pathogen that is not in an ideal environment will slowly die off. Time may be a gardener's best defence against them.

There is a lot written about the importance of killing pathogens with high heat, but consider the source of these pathogens. They are on plant material and soil particles that you collected from the garden. They got to your garden on the air currents from other gardens. Even if you kill all the pathogens in the compost pile, there are still lots in the garden so it is unlikely hot composting will get rid of a disease problem.

Heat Kills Weed Seeds

There seems to be a very high concern about weed seeds in compost. A hot compost pile can kill seeds, but is it really a big concern? Any weed seed in your compost pile probably came from weeds in your garden. If that is the case, is the compost the problem or is it that you are letting weeds go to seed?

A lot of online sources specify a temperature at which seeds are killed, but the truth is that every species is killed at a different temperature and the duration at a higher temperature is also important. For example, from the table below, we can see that 90% of common purslane seed is killed in 1.3 hours at 140°F (60°C), but that it takes 19 hours at 122°F (50°C) to have the same kill rate. But even at the higher temperature, 10% of the seed remains viable.

If weed seeds are a problem, look for a composting method that will kill them, but you may find other gardening techniques that control weeds, like mulching, to be far more effective for controlling weeds.

Number of Hours Required to Kill 90 Percent of Seeds at Various Temperatures.

Weed	Temperature			
	140°F (60°C)	122°F (50°C)	115°F (46°C)	108°F (42°C)
Annual sowthistle	<1.0	2.1	13.3	46.5
Barnyard grass	<1.0	5.4	12.6	unaffected
London rocket	<1.0	4.0	21.4	83.1
Purslane	1.3	18.8	unaffected	unaffected
Black nightshade	2.9	62.0	196.6	340.6
Tumble pigweed	1.1	107.0	268.5	unaffected

Credit: N. Dahlquist et al., 2007

How Long Does It Take?

How long does it take to make finished compost? The answer depends on a number of variables including the input ingredients, the size of the pile, the oxygen level, the moisture level, the air temperature, and how frequently it is turned.

If everything is done correctly, in a backyard setting, it takes about three months in warmer climates and four to six months in colder climates. A cold compost pile can take one to two years.

Commercial operations that use closed vessels can make compost in two weeks followed by a one month maturing process. The first phase of bokashi is done in two weeks, but it is not really composting.

The Carbon to Nitrogen Ratio

You have probably heard about the importance of getting the brown to green ratio correct, but that is mostly a myth. Building a good compost pile is not really about browns and greens; it's about getting the carbon to nitrogen ratio (C/N ratio) correct. People talk about browns and greens because it seems easy for gardeners to understand, but it also leads to a lot of confusion. It is much better to use the C/N ratio.

The C/N ratio is the weight of carbon compared to the weight of nitrogen. The composting process works best when the ratio is about 30:1. This ratio is a mixture of 30 pounds of carbon for every one pound of nitrogen.

Why is this the magic ratio? All organisms require both carbon and nitrogen in their food. They do need other things in their diet as well, but carbon and nitrogen are the key elements. Since microbes are doing the digestion in a compost pile, we are mostly concerned about their diet, and it turns out a 24:1 ratio is a perfect diet for them. The carbon is their energy source, and the nitrogen is critical for making digestive enzymes.

If there is too much carbon in relation to the amount of nitrogen, decomposition is slow because microbes can't find enough nitrogen to consume all the carbon. Such a compost pile never heats up because the microbes never grow to their maximum potential.

If the nitrogen level is too high, the wrong type of microbes prosper and the pile starts to smell. The reason for the smell is that ammonia is given off as a gas, and this reduces the final amount of nitrogen in the compost, which makes it less nutritious for plants.

Why am I recommending a 30:1 ratio when microbes prefer a 24:1 ratio? The reason is that the C/N ratio drops during composting as carbon is lost as CO_2. A pile that starts out with a 30:1 ratio ends up with a ratio closer to 15:1. By starting with some excess carbon, we keep the microbes happy for a longer period during the process.

All organic matter contains both carbon and nitrogen, but each type of material has a different ratio. Wood has a very high ratio—a lot of carbon and very little nitrogen. Manures have a low ratio. Green leaves have more nitrogen than brown ones in fall. The C/N ratio of various input ingredients are given in the table.

It is important to understand that the values listed in the table below are approximate values. Cow manure with more urine in it has a higher nitrogen level than the same manure with less urine. Older manure also has less nitrogen than fresh manure. This is one reason you should never use fresh chicken manure near plants—it contains a lot of nitrogen and can burn the plants. Aged chicken manure has much less nitrogen and is safe for plants.

Getting the Right C/N Ratio

The goal is to make a pile with a C/N ratio of 30:1, but how do you make such a pile? The process of calculating all of this is known as a compost recipe.

I am going to describe a couple of ways to work out the recipe. The first one (Accurate Method) is more complicated, but understanding it will give you some valuable insight into the process. The second method (Easier Method) is less complicated, more practical, but not as accurate, and the third method (Even Easier Method) is really easy to use.

Organic Materials

Material	Brown/ green	C/N ratio	% Carbon	% Nitrogen
Alfalfa pellets	green	15	40.5	2.7
Blood meal	green	3	43	13
Coffee grounds	green	20–25	25	1
Cottonseed meal	green	7	42	6
Fall leaves	brown	30–80	20–35	0.4–1.0
Fruit waste	green	15–35	8	0.5
Grass clippings, fresh	green	15–25	10–15	1–2
Hay, dry	green	20–25	40	1.5
Kitchen waste	green	15–20	10–20	1–2
Manure – chicken	green	6–15	15–30	1.5–3.0
Manure – cow	green	20	12–20	0.6–1.0
Manure – horse	green	25–40	20–35	0.5–1.0
Newspaper, cardboard, dry	brown	200–500	40	0.1
Seaweed, fresh	green	10–20	10	1
Soybean meal	green	7	42	6
Straw, dry	brown	60–100	48	0.5
Vegetable waste – leafy	green	10–15	10	1
Vegetable waste – starchy	green	15	15	1
Weeds, fresh	green	10–30	10–20	1–4
Wood chips, sawdust	brown	100–500	25–50	0.1
Yard waste	green	30	40	1.3

Note: Carbon and nitrogen are found in compostable material. Percent values are based on fresh weight.

Compost Myth: The Ratio of Browns and Greens

The common advice for making compost is that you should use a 30:1 ratio of browns to greens. In the simplest form, the terms are quite descriptive. Browns are any plant material that is brown, including fall leaves, dried grass, wood products, paper, and straw. Greens are—you guessed it—green. This category includes fresh grass clippings, freshly picked weeds, plant clippings, and most kitchen scraps.

Calling composting ingredients brown or green is useful because it is simple for people to understand. However, the terms are not always correct. It would be better to use the terms high nitrogen ingredient and low nitrogen ingredient. The greens contain higher levels of nitrogen. For example, fresh green plant material contains high levels of nitrogen. As the greens age and turn brown, they lose nitrogen. Green leaves have high levels of nitrogen, but as they go brown in fall, the nitrogen level drops.

So is manure a brown or a green? Based on color, it is a brown, but based on nitrogen levels, it is a green. As far as composting goes, manure is a green. Other ingredients are also confusing. Alfalfa hay is brown in color, but it is also a "green" since it contains a lot of nitrogen.

The second problem with the above advice is that the ratio of 30:1 applies to carbon and nitrogen and not to browns and greens. When dealing with browns and greens, the ratio is closer to 1:1.

The brown to green ratio is easier to use, but it is very approximate. Browns can be green, greens can be brown—it gets confusing! Overall, it is better to stick with the carbon to nitrogen ratio.

A lot of gardening advice will tell you to use the C/N ratio of each item and figure out how much of each ingredient you should add, but that is a very complicated process. One problem is that you don't know the moisture content of each item, and water can add significant weight to any item. Fresh-cut grass is much heavier than dry grass, and this weight has to be accounted for.

You also have to account for the weight of chemically bound oxygen and hydrogen in the material. For example, starch is 44% carbon, 6% hydrogen, and 49% oxygen by weight. The weight of the oxygen is significant.

Although the C/N ratio is heavily promoted in garden circles, it is really too complicated to use unless you have a degree in mathematics. What works much better is using the percent of carbon and the percent of nitrogen, on a fresh weight basis. If you use these values, you can ignore the weight of the water, hydrogen, and oxygen.

An Accurate Method

Start by getting your ingredients together and weigh each material. Look up the percent carbon and percent nitrogen of each one, in the above table. If you can't find your material, pick something that is close to it.

Calculate the weight of carbon and nitrogen in each item.

Example:

If you have 3 pounds of coffee grounds, you have 3×0.25 (25%) = 0.75 pounds carbon. You also have 3×0.01 (1%) = 0.03 pounds of nitrogen.

Next, calculate the total weight of carbon by adding up all of the carbon values. Do the same to get a total nitrogen weight.

The C/N ratio is calculated as follow:

C/N ratio = weight of carbon/weight of nitrogen.

It is unlikely that you end up with a perfect C/N ratio, but if you are between 20 and 40, you are close enough. Go ahead and make your pile.

If the calculated C/N ratio is too high, you have too much carbon so remove some of the high ratio material or add in some more low ratio material until you have a C/N ratio between 20 and 40.

If the calculated C/N ratio is too low, do the opposite.

This method is quite accurate but requires more calculations. It is a good method to use if you want to make really fast hot compost.

An Easier Method

One of the problems with the above method is that it requires you to weigh your materials. The reason for this is that weights are much more accurate than volume because you can ignore the bulk density of items. But gardeners work in volumes. They use buckets, shovels, and wheelbarrows, not scales. And, let's face it, many don't want to do a lot of complex math.

Take your input ingredients and make three piles. Pile A is high nitrogen material and includes anything with a C/N ratio of 30 or less. Pile B is the medium-high carbon pile and includes anything that has a C/N ratio between 30 and 100. Pile C contains anything with a C/N ratio above 100.

Using a pail, shovel, or wheelbarrow, add two parts of pile A for every one part of pile B. Then add eight parts of pile A for every one part of pile C. Mix it all together in a single pile.

Watch the pile. If it starts to smell like ammonia, you added too much nitrogen so add more high carbon material (pile B or C). If it is not heating up, add more high nitrogen material (pile A).

This method is much easier than the Accurate Method. It will produce compost, but you may not get a really hot pile.

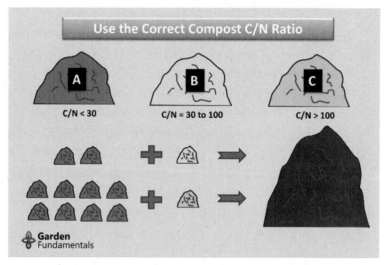

Easy method for getting the correct C/N ratio.

An Even Easier Method

If you like to keep things really simple, mix browns and greens in a 1:1 ratio, based on volume.

It will also give you compost.

Cold vs Medium vs Hot

So far in the book, I have discussed hot composting, and that seems to be the goal of many new composters. Everyone wants to do it right, get it hot, and make it fast. There is nothing wrong with that goal, but medium or cold compost might suit you better, and they are both great for the garden.

Fast compost is made with a C/N ratio of 30:1. When this ratio is in the range of 40 to 200, the pile does not heat up as much and composting is slower, taking three to nine months to finish. If the ratio is even higher, above 200, there is no perceptual heat and it can take a couple of years.

Compost Myth: pH Is Important and Should be Monitored

A lot of discussions about composting talk about the importance of pH. Don't use acidic ingredients because it makes the compost too acidic. Measure the pH during the process so you can keep it in the right range.

The reality is that pH is not that important.

A lot of the so-called acidic ingredients like pine needles and coffee grounds are not even acidic. That is another common myth.

Most of the organic material that a homeowner would use has a neutral pH around 7.0. Input ingredients do affect compost pH, but these differences tend to be small. A study that looked at homemade compost found pH values between 7.0 and 7.5. The pH will drop a bit as compost sits or is used in the garden.

You don't have to monitor pH, and you don't have to worry about compost changing the pH of your soil. There is also no need to add lime to a compost pile—that is another myth.

The key point to understand is that all three of these will produce good compost.

Everyone talks about hot composting, but the reality is that 90% of backyard compost piles are medium piles. There is some heat, but not a lot. The process moves along, but not too quickly. This is even truer if the pile is not turned, which is quite common. This type of composting is less work, and you don't have to worry as much about getting the C/N ratio right.

I will also discuss other composting systems like rotating drums and small plastic compost containers. All of these are medium or even cold systems.

Chemical Properties of Compost

The pH of finished compost depends very much on the materials used to make it. If you use wood products like sawdust, it will make the finished compost more acidic. If you use more manure or add in some ashes from the fireplace, it will be more alkaline.

As the compost is being made, it goes through pH swings. In the initial stages, organic acids are formed. These make the compost pile more acidic and the pH drops. In these acidic conditions, fungi grow better than bacteria, and they take over the pile and start to decompose the lignin and cellulose. As a result of this, the pH rises and bacteria become more populous. What this means is that the

The pH of Some Common Composts

Compost Type	pH
Yard debris	7.7
Mixed manures	7.9
Leaf mold	7.2
Horse manure	6.4
Bark	5.4
Homemade garden compost	7.0 to 7.5

pH of your finished compost also depends on when you consider it to be finished. If you rush things and use it early, it will be more acidic. If you wait longer, it will be more alkaline.

The nutrients in compost are relatively low compared to synthetic fertilizers. Home compost was found to have an average NPK of 3-1-1.5. Many commercial compost products have an NPK of 1-1-1.

Loss of Plant Nutrients

The way I have described the composting process so far would lead you to believe that the finished compost contains all of the plant nutrients that were in the starting ingredients, but that is not the case.

Nitrogen is the most important plant nutrient, and it is the one that is most commonly deficient in soil (discussed in my book *Soil Science for Gardeners*). It is also the nutrient that is most easily lost during the composting process.

Any free urea (animal urine, or fertilizer) or ammonium in the pile is easily lost to the air as it heats up. Most of the nitrogen is bound up in large molecules inside the organic matter, but as it is released, it becomes very soluble and some of it turns into ammonia gas. Excess moisture or rain easily washes the soluble nitrogen out of the pile, and that is one reason for covering compost piles. As much as 30% to 50% of the nitrogen in the input ingredients can be lost, depending on the starting material and the composting process. Turning the pile frequently increases the loss.

Potassium is a nutrient that remains in ionic form inside plant matter. As the cell structure in the material is broken down, it is easily leached out of the pile.

Phosphorus in organic matter is available as simple molecules and as large molecules. During decomposition, it is slowly released, but since phosphate is not very soluble in water, much of it will remain attached to the finished compost. This is one reason why the finished compost NPK tends to have a higher middle number.

The above discussion about the release of nutrients is only part of the story. A large proportion of the nutrients remain in the form of large molecules and are released very slowly over many years.

Finished compost will, on average, release nutrients over a five-year period. That is one of the key benefits of compost.

Humus

Gardeners use the term "humus" in a number of confusing ways. The term is used to describe good soil, as in humusy soil, and many suppliers stick the term on their product to increase sales. Soil is not humus. The term is also used in place of "finished compost." Finished compost is not humus nor does it contain a lot of humus. It might be 10% humus.

Even scientists have had difficulty defining the term. About 200 years ago, soil scientists noticed that good agricultural soil was black, and in an effort to better understand this black stuff, they devised a method to extract it. They treated soil with a strong alkaline solution of pH 13 that pulled the organic component out of soil so they could study it. Over time, this black substance became known as humus.

Humus has some very unusual properties. It is composed of mostly carbon and some nitrogen, which is not unexpected. The unusual part is that humus is very stable. In fact it takes 100 years or more for it to decompose. Microbes can't seem to digest it even though microbes can digest just about every other organic material, including oil.

It was thought that as compost continues to decompose over time, it becomes this stable humus. Once it is humus, it still improves the structure of soil but it no longer releases nutrients for plants.

There are two important points in this story. The first is that in over 200 years of study, nobody has been able to extract humus from soil except at very high pH. The problem is that such a high pH changes all kinds of chemical structures. So humus may not actually exist in nature.

The second issue is that the molecules in humus are huge. The decomposition process is one where large molecules are broken down into smaller and smaller molecules. And then suddenly, we have these huge humus molecules. No one has ever been able to explain this or show that it actually happens in soil.

Today scientists believe that humus does not really exist. The treatment of soil with high pH solutions causes the formation of large humus molecules. They don't actually exist in soil. The stuff that is in soil is organic matter that is constantly changing. Decomposition processes break down organic matter into smaller and smaller molecules, but at the same time, microbes and macroorganisms take these small molecules and turn them into large ones again, a process called immobilization. It is a never-ending cycle that benefits all soil life including plants.

It is more correct to call this material "organic matter" rather than "humus."

When Is Compost Ready?

This question can be broken into two different questions: when is compost finished and when can you use compost?

When Is Compost Finished?

As the compost pile reaches the end of the process, the material becomes black and unrecognizable. The tomato plant and the banana peel are no longer there, but some of the hard stuff like twigs and walnut shells may still be recognizable. The material feels cool and has an earthy smell. This is what gardeners call finished compost.

But this compost is not finished. It will continue to decompose for several more years.

Scientists have also been trying to define finished compost, and they haven't been able to do it very well. There are several lab tests for measuring the completeness of the process, but none of these are perfect. One of the more accepted definitions for finished compost is compost in which seedlings can grow. You can easily carry out this test yourself, as described in the sidebar for the Seedling Test.

If the Seedling Test is successful, the compost is finished. If not, allow it to age longer.

Seedling Test for Compost

The Seedling Test is a great way to evaluate compost to make sure it will not harm plants.

Take some compost, mix it with 50% soil, and put it in a pot. The size of the pot is not important, and any garden or potting soil will work. This is your test pot. Take the same size pot and fill it with just soil to create a control pot.

Add the same number of easy to germinate seeds to both pots; radishes, lettuce, beans, or peas work well.

Water the pots and wait for the seedlings to grow. Compare the test pot to the control pot. If the plants grow properly and have no deformities, there is nothing wrong with the compost and it can be used in the garden or in containers.

If the test seeds don't sprout or are deformed, the compost should not be used.

If this is fresh compost, the problem could be that it is just too fresh. Let it age a while and test it again.

The compost could also contain harmful chemicals like herbicides. In that case, aging it will probably not resolve the issue. Such compost should not be used with seeds or young plants. It may be suitable as mulch around established older plants.

When Can You Use Compost?

The answer to this depends very much on how you want to use the compost.

As mulch, you can use it any time and it doesn't need to be finished. In fact it doesn't even need to be composted. Collected leaves in fall can be used as mulch. The cut and drop method discussed later in the book can be used as mulch with no composting.

Since mulch sits on top of the soil and does not come in contact with plant roots, it won't harm plants.

Compost that will be buried in soil can also be unfinished, depending on when you add it and when you plant. If you want to add it to your garden in fall and plant in spring, it does not need to be finished because the finishing process will happen over winter right in the soil. If you dig it into soil right before planting, it is safer to use finished compost.

Some people will dig it into soil right before planting but only put it in unused rows between rows of plants. That way the roots don't come in direct contact, for a while, allowing you to use unfinished compost.

When compost is mixed with soil and used in containers or for houseplants, it is best to use finished compost.

4

Compostable Material

What can you compost? The short answer is that you can compost anything that is plant or animal based. In agriculture people even compost dead cows. There are also a number of controversial ingredients like pet waste or even human waste. You can certainly compost these and many do, but you may not want to.

A more appropriate question is what do you *want* to compost? Some items decompose very slowly, and you might decide to leave them out. Others, like pet waste, pose potential health risks. Some smell and therefore attract animals which you may or may not want around your garden. I don't have to worry about bears, but they are a big concern in some areas.

This chapter lists many things that you can compost, and to organize the list, I have broken it into three sections: the Good, the Bad, and the Controversial. The good are things that most people compost. The bad are things you should not use, and the controversial, well, they're controversial.

Keep in mind that the material you use also depends a bit on the composting method. Composting a cow works well in a windrow (see windrow composting in Chapter 6) but not in a backyard tumbler.

Good Composting Material

Animal Manure

Animal manures are ideal for making compost. They are a waste product, and in some locations, they are readily available. Before we became so fancy about composting and started making fertilizer, manure was the go-to material for developing healthy soil. It contains plant material that is already chopped up for you as well as higher levels of nitrogen.

There are a few concerns with it.

Pathogens, such as *E. coli* and salmonella, can exist in manure, and they can cause health issues. Hot composting and bokashi are normally recommended for manure because they kill off the pathogens. Pathogens die off over time, so even cold composting works if it is done for a longer period of time.

There is also a concern with the antibiotics used to treat animals, but manure from organic farms will not contain these. There is evidence that plants can accumulate antibiotics from soil amended with animal manure, but this is not considered to be a health risk. If you are concerned, use such compost on ornamental gardens where it is not an issue. Personally, I feel that people have gotten overly concerned about such chemicals in their food. After all, even plants make antimicrobial compounds, and we don't worry about eating them.

The addition of antibiotics to the environment is a valid concern because it can lead to resistant mutations.

Farm animals eat plant material that contains seeds, and some of these seeds pass through the animal. Manure can contain significant amounts of seeds, especially from horses. These are killed off with hot composting.

In some cases, herbicides are used on grain crops to control weeds. When these plants are fed to animals, the herbicide survives the digestion process and ends up in the manure. Numerous cases have now been reported in both Europe and North America where commercially composted manures harm plants because of high levels of herbicides, like aminopyralid.[1] This is not a problem with the glyphosate found in Roundup.[2]

Commercial composting is normally well controlled and hot. If their process does not eliminate the herbicide, yours won't either.

Should you use manure? The answer is a definite yes, but there is value in knowing your source, and if you are not sure, do a Seedling Test as described earlier. It will reveal existing herbicide issues.

Coffee Grounds

Coffee grounds are claimed to have high nutrient values, but that is not really true. They contain 1–2% nitrogen, 0.3% phosphorous, and 0.3% potassium, along with a variety of micronutrients. Their C/N ratio is 23:1. That means they are not high in nitrogen and have an almost perfect ratio for composting.

Contrary to popular belief, they are not acidic, so they won't make compost acidic.

Adding too much coffee grounds directly to a garden can harm seedlings due to high levels of caffeine, so it is best to compost them first.

Eggshells

Lots of people add eggshells to their compost pile and feel that they have added calcium to soil. But that is mostly a myth.

Eggshells do not decompose in most soils or in a compost pile to any great extent. The inside of the shells have an inner protein skin, and this decomposes quickly. What is left is hard calcium carbonate.

I performed a study to show how fast eggshells decompose. I buried some in my garden, and in each of the following five years, I dug them up and examined them for degradation. There wasn't any. Eggshells do decompose very slowly in acidic conditions where the pH is below 5.0, but only if they are highly pulverized. There may be some limited degradation in the early composting phase as the pH drops, but experienced gardeners know that eggshells survive the composting process.

There may be more degradation in the acidic conditions found in bokashi and eco-enzymes.

Food Scraps

Food scraps, both raw and cooked, are excellent for composting and most contribute extra nitrogen. If it is plant-based, go ahead and add it to the pile. However, there are some items that can cause a problem and are usually not composted in backyard situations.

Meat and fish can attract animals. That does not harm the compost, but you may not want to attract rats, dogs, and bears from the neighborhood. Methods such as bokashi are more suited to composting meat.

Bones do compost, but very slowly and add little value to your compost.

Oils, fats, and greases do compost, but they can cause problems in larger amounts, and it is best not to use any significant amounts. A little oil on your leftover salad is not a problem, but a pail full of grease creates an anaerobic environment and kills off a lot of the good microbes.

Dairy products are usually not composted because they attract

Compost Myth: Don't Compost Old Potting Soil

Should you compost old potting soil?

The concern with old potting soil is that it contains plant diseases, and by adding it to a compost pile, you might spread them all over the garden or to other houseplants. There is a slight risk of this happening, but it is unlikely. Any disease organism on houseplants has probably already spread to other plants. A disease in the garden from container plants is also not a concern because those diseases are already in the garden.

One thing to understand is that potting media does not add a lot of value to a compost pile. Any soil in it does not help with composting. The soilless part of the mix will decompose, but things like peat moss or coir add very few nutrients. If you want to get rid of it, just dump it in the garden.

What should you do with old potting media? Reuse it. There is a common belief that this material needs to be replaced on a regular basis. That is a myth. It can be reused for many years. If you have hard water, scrape the top one inch off to remove any white deposited salts and reuse the rest. If it has settled, top it up with some fresh material and plant in it.

You can compost potting material, but there is really no reason for doing it.

animals, but other than that there is no reason not to add them, provided that you don't add too much. Liquid items like milk are best mixed with dry ingredients such as leaves or sawdust, or just pour it directly on your soil.

Large seeds from things like peaches and avocados decompose very slowly, but they can be added to soil after composting and will eventually degrade.

Citrus rinds from things like oranges and lemons have antimicrobial properties, but they will compost eventually. Some experts say that they should not be fed to worms in a vermicomposting system, but others disagree and say they are fine in moderation, and in moderation, they won't harm the worms.

Grass Clippings

Grass clippings are excellent material, and when green, they contain a lot of nitrogen. The problem with them is that too much fresh material releases a lot of water and nitrogen, which turns into a smelly anaerobic pile.

You can dry the grass before adding it to prevent the smell, but it still contains a fair amount of nitrogen so mix it with high carbon material.

The best place for grass clippings is right on the lawn. They don't cause thatch as is commonly believed, but they do provide about 25% of the annual nitrogen needed by the lawn. Use a mulching mower and leave the clippings where they fall.

Lawn clippings can also contain herbicide. Composting reduces the amount of herbicide but may not eliminate it. You can still use such compost as mulch around trees, or on the lawn, but it is best to keep it out of the garden. You can do the Seedling Test to check for an herbicide problem.

Hard to Compost Items

Some items are very slow to compost: corncobs, ornamental grass stocks, and peanut shells, for example, but there is nothing wrong with adding them.

Hay or Straw

Hay and straw are both good materials to add to a compost pile. They don't pack down as much as some other material, which provides good air exchange. They are high in carbon, and so they do need to be balanced with a good nitrogen source.

Since most people have to buy hay and straw, I think it is better to use it as mulch in the vegetable garden. Let nature do the composting over time.

Hedge Trimmings, Branches

Woody branches will decompose, but it happens very slowly. Some people put a layer of them at the bottom of a compost pile to increase air flow.

In my experience, they decompose too slowly even in a hot pile, so I don't add woody stems to any of my compost systems. Instead I pile them up to make homes for wild animals and hiding places for birds.

Leaves

Leaves are excellent material for composting. They do tend to mat down so it is better to chop them up, or mix them with other loose material.

Paper Products (Cardboard, Newsprint, Napkins, Paper Towels)

There is no simple answer to this one. It really depends on the type of paper product. There are some general points to consider.

1. All paper products decompose very slowly.
2. There are almost no nutrients in paper so all they add to compost is some carbon. The idea that "worms love eating cardboard" is false—there are no nutrients in paper for them.
3. Some paper products contain toxic materials.

Most newspaper, napkins, paper towels, and cardboard are safe to compost. The inks used on them are usually plant-based dyes that are not toxic. The general public has raised the following concerns about composting paper.

Bisphenol A (BPA)

Some people have raised a concern about using paper products because they use some recycled paper and this can contain a toxin called Bisphenol A (BPA), mostly from recycled thermal cash register receipts.

A study[3] looked at 15 types of paper products including thermal receipts, flyers, magazines, newspapers, food contact papers, food cartons, printing papers, and paper towels and found that the exposure of the general population to BPA from these products is "minor compared with exposure through diet." BPA is used for lining metal cans, in plastic wraps and polycarbonate plastics, and is found in a number of food products "including fresh turkey, canned green beans, and canned infant formula."

BPA exposure is a concern in other areas of our life—but exposure through composted paper is not an issue.

Chlorine and Dioxin

Bleached paper such as napkins, coffee filters, toilet paper, and paper towels is white in color and can contain chlorine and dioxin (a known carcinogen). The chlorine levels are not a concern since they exist at very low levels and some chlorine is needed by plants.

Several studies by both the EPA and the Ministry of Health in New Zealand have shown that the levels of dioxin are so low they do not cause a health risk. Dioxin degrades quickly when exposed to sun, and its half-life on the surface of soil is one to three years. Dioxin does not seem to be a big problem with composted paper, provided you are not composting huge amounts.

Glue in Cardboard

Glue in cardboard is also a concern. There are two places where glue is used. One is to make the actual cardboard and the second is to form the boxes.

The glue used to make cardboard is almost exclusively made from starch which is derived from natural carbohydrates found in roots, tubers, and seeds of higher plants such as maize, potatoes, wheat, rice, and tapioca. They easily degrade in the composting process.

Glue is also used to form the boxes, and it's less clear which glue is used in any particular instance. However, the amount of this glue is minimal.

Glue on cardboard is not a real problem.

Ink and Heavy Metals

There are two classes of inks used to print paper and cardboard: vegetable dyes and colored inks.

Most newspaper and non-glossy paper use vegetable dyes. These are perfectly safe in the garden.

Colored inks are used on some glossy paper and some cardboard, like cereal boxes. The problem with these inks is not the ink itself but the fact that they may contain heavy metals. Heavy metals do not decompose in a compost pile or in soil. Plants do absorb them from the soil, and both plants and animals accumulate them in tissues, which mean our bodies have more and more each year. Even quite small amounts of heavy metals are a health concern.

One thing to remember is that native organic soil also contains heavy metals. A Canadian study[4] looking at soil samples found average lead values ranging from 13 to 750 mg/kg, but this can be higher in older neighborhoods and industrial areas. Compare that to 2.6 mg/kg found in recycled cardboard. You would need to add a lot of composted cardboard to make a significant change to most soils.

Also consider that bringing any type of organic material into your garden also adds heavy metals. This includes manure, compost, and mulch. Plants accumulate heavy metals, and bringing them onto your property increases the metals in your soil.

An interesting study[5] showed that adding compost made from biosolids (sewage sludge) reduced the amount of lead absorbed by plants. Organic matter has a high cation exchange capacity (CEC) and holds on to heavy metals, which prevents roots from getting to them. Adding compost, even if it adds some lead, results in less lead being absorbed by plants.

Adding composted paper containing heavy metals is not a great idea, but small amounts won't impact the heavy metals in the food you produce.

Should You Compost Paper or Cardboard?

Understand that paper and cardboard composts very slowly because of its high lignin content. Personally, I found that even shredded paper is still mostly intact when the rest of the compost is done. I don't think it is a good addition to a compost pile.

Is it safe? White non-glossy paper, like newspaper, or office paper is quite safe. Dioxin, dyes, chlorine, and BPA are not a big concern.

Any composted paper is safe in a non-food-producing garden.

Printed glossy paper and cardboard contain low amounts of heavy metals which could be a concern. Unless you use a lot, you probably get a higher dose of heavy metals from driving to work (smog, exhaust, tire dust, etc.) than from eating produce from your garden. But heavy metals accumulate in your body, so it is prudent to try and keep the level low.

Pine Needles

Contrary to popular belief, pine needles are not acidic. I would not use them in most methods except for a hot compost pile, and even then they decompose very slowly.

They are a great mulch, and I think that is a much better use for them.

Sawdust, Wood Shavings

These are very high in carbon and decompose very slowly. They are great for adding to other high nitrogen material like manure. If your input ingredients are mostly kitchen scraps and yard waste, you will have a hard time finding enough nitrogen sources to mix with wood products.

Small amounts can be added without any issues.

Seaweed

Seaweed is organic material and works well in a compost pile, but there are some potential issues with it.

It can contain a lot of salt which can be toxic to the microbes and the plants in your garden. Provided the seaweed is washed to remove the salt, it can be used.

The second issue is the way in which it is collected. Collecting material that has washed up on beaches is fine. Harvesting it from the ocean is not.

Sod

Sod is removed from a lawn when you make a new garden, and some people add it to the compost pile. This can work fine but try not to add too much soil along with the sod.

Controversial Composting Material

Diseased Plant Material

A lot of sources say that you should not compost plant material that is diseased. I have news for you. Virtually all plant material in the fall is diseased. If you don't use it, there is very little to compost.

Learn to understand specific diseases. For example, powdery mildew is in almost every garden, and you probably recognize it. But did you know that just about every species of plant is infected with a different species of powdery mildew? The powdery mildew on your roses won't infect your lilacs.

The other thing to understand is that fungal spores are almost everywhere. They are on plant leaves, plant stems, and they cover the soil. They also travel great distances to find your garden. You can't keep them out.

Some maples get large black spots, a fungal disease called tar spots. Some people won't compost their leaves because they want to prevent the disease in future years. Tar spot fungal spores are all over your city. No matter what you do on your property, your maples will be infected again next year. By the way, the spots are mostly an aesthetic issue, and they do very little damage to the plant.

Removing infected leaves from your garden will not prevent most diseases, so you might as well compost them and get some value out of them.

There are exceptions and this is why it is important to understand specific diseases. Some very contagious and harmful diseases can be controlled by getting the diseased material off your property. For example black knot on the prunus family (cherries, chokecherries, and plums) should be removed from the property as soon as you see it.

Compost Myth: Some Plastic Is Compostable

Some plastic is compostable, but that does not mean it will decompose in a compost pile.

Compostable plastic is defined by the standards association, ASTM International (ASTM), as "a plastic that undergoes degradation by biological processes during composting to yield carbon dioxide (CO_2), water, inorganic compounds, and biomass at a rate consistent with other known compostable materials and that leaves no visible, distinguishable, or toxic residue."

Note that neither the time frame nor the conditions for composting are included in the definition.

Many assume that all compostable plastic is a new type of plant-based plastic. Some compostable plastic is made from plant material such as corn, potato, tapioca, soy protein, and lactic acid, but others are made from petroleum, including BASF's popular Ecoflex product.

Most compostable plastic requires a high temperature for an extended period of time. For example, polylactic acid (PLA) is a popular compostable plastic used to make drinking cups, clamshell containers, and plastic cutlery. It requires 140°F (60°C) for 60 to 90 days to decompose. Even most municipal operations do not meet that requirement, let alone backyard composters.

Compostable plastic does not compost in your backyard, no matter which method you use.

Many tea bags are PLA made from corn and are promoted as being 100% compostable, but if you read the fine print, they won't compost in a home composter.

What Is Better—Regular Plastic or Compostable Plastic?

This is a good example of the value of science.

Plastic made from plants sounds so organic, and when they are labeled with "compostable," they become irresistible to the public, who imagine plastic drinking cups melting away in a compost bin. Many manufacturers are quick to reinforce that vision, and the media also jumps on the bandwagon with headlines like "Calgary Co-op to eliminate plastic bags from liquor stores as compostable bag program takes off."

Now science steps in and asks which option is better for the environment? Testing is still going on, but it appears that given our current composting capabilities in North America, traditional plastic might be more eco-friendly because compostable plastic can't be composted with current facilities, whereas regular plastic can be recycled.

Ground Stone (marble) and Shells

Things like rock dust and ground shells decompose very slowly and only during the initial composting phase when the pH is low. They are not a useful addition to compost.

Human Waste

There are two components to consider: solid waste and urine.

Urine contains 0.9% urea, or about 0.4% nitrogen. It also contains 0.1% potassium, so it has an NPK of 0.4-0-0.1. Unfortunately it also contains 0.2% sodium which can be toxic to plants and microbes at higher levels.

Urine can be added to your compost pile as is or applied directly to the garden after diluting it by a factor of ten.

Human feces are more controversial. There is the ick factor which will keep most of you from using it, but consider that in many developing countries, it is called "night soil," and it plays a major role in fertilizing crops.

A major concern with feces is the transmission of diseases and pests. For example, most giardia infections are the result of exposure to feces from someone already infected. This is a controversial subject, and if you want to understand it better, have a look at *Essential Composting Toilets*, by Gord Baird and Ann Baird.

A significant problem with using feces is that we don't have easy-to-use collection systems for it.

Pet Waste

There are 78 million dogs in the US, producing 10 million tons of waste each year. That is enough to fill a line of tractor trailers lined up end to end from Boston to Seattle. The average dog produces 400 pounds (180 kg) of waste per year, and it has an NPK of 2-10-0.3. That is as much nitrogen as you get from an 18-pound bag of urea, but the dog waste also adds organic matter.

Most authorities on composting and health experts will tell you not to include pet waste. I disagree, at least in some cases.

Pets carry parasites and diseases which are collectively called zoonotic organisms. Their feces can transfer these to soil directly or through composted waste, and there is a risk that you can get sick from that soil. But you can also easily catch zoonotic diseases from their fur, from footprints on the carpet, and even from walking your dog down the sidewalk, since dogs tend to spread small amounts of their poop on the ground as they walk.

But here is a point many miss. If the pet lives with you and you handle it, you have a much greater risk of getting sick directly from your pet than from compost made using their waste. So you might as well compost their waste.

It is less advisable to compost other people's pet waste unless you use hot composting. A temperature of 145°F (63°C) for a short while, or 130°F (55°C) for a minimum of three days, will kill worms, worm eggs, and diseases. The risk in cold composting is very low, but it does exist. For more information on this topic visit https://www.gardenmyths.com/compost-dog-cat-waste/.

The worms in vermicomposting do a good job digesting the poo, but there is no guarantee that parasites are killed in the process.

Cat waste is not that different from dog waste. It can also be composted, and it also contains organisms that cause zoonotic diseases. The difference is that most cat waste is combined with kitty litter which commonly consists of different types of clay. The clay won't compost, but it does not harm the compost pile. The clay may not be great for soil that already has a high clay level, but it can be beneficial for sandy soil.

Rhubarb Leaves

Rhubarb leaves are reported to contain high toxic levels of oxalic acid, and for that reason, they are not recommended for composting.

I had a close look at this in my book *Garden Myths Book 2* and found that the levels of oxalic acid in rhubarb leaves is about the same as in carrots and radishes, and less than found in parsley or spinach. Rhubarb leaves are perfectly safe for composting.

Toxic Plants

Some gardeners are concerned about composting toxic plants because they don't want toxic chemicals in their soil. Why? These toxic compounds might be taken up by food crops rendering them toxic.

This all seems logical, but it is a myth.

First of all, all plants, including the plants and fruits you eat, contain toxic chemicals; 99.9% of the pesticides you consume are natural pesticides made by the plants to defend themselves from predation. The tomato plant you have been composting contains toxic alkaloids called solanine and tomatine, which by the way are also found in the fruit you eat, especially in green tomatoes.

The second key point is that the dose makes the poison. There are not enough alkaloids in a tomato to harm you. Now consider what happens when you put that tomato in a compost pile and dilute the toxin by all of the other stuff in the pile? As the dose goes down, the risk also goes down.

The third point is that all of these toxic compounds are organic in nature, and microbes digest them. As each day goes by, there are fewer and fewer of them in the compost and in the soil where compost was spread.

Toxic plants can all be composted without any concern for toxicity.

Weeds (*pernicious*)

Should you compost weeds? Most gardening advice will tell you to keep weeds out of your compost pile so that you don't spread them around your garden. I disagree.

The first issue with weeds are weed seeds. If they get in the compost pile and are not killed during the process, they will be spread around your garden as you distribute the compost. The cure is to control weeds. Don't let weeds flower, and if they do, don't let them make seed. This is just smart gardening even if you don't compost. If you follow this advice, weed seeds are no longer an issue.

I group weeds into two categories: easy to control weeds and pernicious weeds.

Easy to control weeds are those that don't have aggressive runners, or other vegetative ways to spread. They are also ones that are relatively easy to pull out. These weeds can be composted and are not a concern for spreading because composting will kill them.

Pernicious weeds like Canada thistle, goutweed, and bindweed are more of a concern. They spread by runners, and any small bit of runner will grow into a new plant. Adding these as living plants could help spread them, especially in colder composting systems.

There are a few options for handling this group. I like to lay them out in the sun for several days so they are good and dead before tossing them into the compost pile. You can also lay them on a lawn to dry and let them compost there.

You can also put them in a pail of water and let them ferment for a week. Then add them to the compost pile. Doing outdoor bokashi will also kill them.

And then there is Japanese knotweed that makes thick tubers and is very hard to control, or running bamboo that has runners which seem impossible to kill. If you have really nasty weeds, put them in the garbage but not the green recycling bin.

Wood Ash

Wood ash has an NPK of 0-1-3 and a calcium carbonate level of 20%. People add ash to their pile because of the potassium, and that may be a valid reason if your garden needs more potassium.

The problem with wood ash is that the pH is between 10 and 12, which is quite high especially if you have alkaline soil.

Ash does not compost, so there is no point in adding it to a compost pile. If your soil is acidic and you want to increase the pH, add it directly to the soil.

Ash from barbecue briquettes should not be used anywhere in the garden.

Bad Composting Material

Disposable Diapers

You can compost human waste and paper, why not diapers? First of all, disposable diapers contain plastic which does not decompose, but more importantly, they contain hydrogels.

The absorbing property in diapers is due to hydrogels which are cross-linked polyacrylamide polymers. These are environmentally safe, but over time they degrade to acrylamide which is a known neurotoxin and suspected carcinogen. It is not something I would add to my garden soil.

Roses and Raspberries

I have picked on roses and raspberries, but this comment applies to all woody plants with thorns. They can be composted, but since they are woody, they take a long time to decompose, which means your finished compost is full of thorns.

I hate thorns in the garden!

My roses get cut back in spring, and all the cuttings go in the green recycling bin. I know that if I leave them in the garden, I will stick myself with them.

Sand and Soil

Adding sand and soil will do no harm, but they don't help either. They just increase the weight of the compost, making it harder to move around.

The one exception is a few handfuls of soil that are added as you build the pile to add extra microbes. This is completely unnecessary but it does give some people comfort, knowing that they added extra microbes.

More Bad Composting Items

Here are some more items that you should leave out of your compost pile: carpeting, dryer lint, fabrics, hair, and vacuum cleaner contents.

Compost Myth: Comfrey Is a Dynamic Accumulator

Are you looking for more green material for your compost bin? Why not grow it?

There is a lot of interest in growing high nutrient plants for the compost pile, and comfrey is at the top of that list. The reason is twofold: it grows fast, creating a lot of material, and it's claimed to be a dynamic accumulator.

Comfrey does grow fast and produces a lot of vegetation. If you want to grow more greens for your compost pile, then this is a good choice.

Unfortunately, the second reason for using it is a myth.

A dynamic accumulator is a plant that has much higher nutrition than other plants, and it gets those nutrients from deep in the soil using a large taproot. Comfrey meets neither of these conditions. It is not particularly high in nutrients, and most of its roots are near the surface. It doesn't really form a taproot either.

Plants do not get their nutrients from deep in the ground as some believe. The nutrients, air, and water that plant roots need are near the surface of the soil, and that is where most roots develop.

5

Managing the Composting Process

Making compost is part art and part science. There are some basic principles that need to be followed, but there are also many variations on each type of composting. Gardeners love to experiment and add their own twist to procedures.

It is also important to understand that many gardening procedures and rules have not been examined by scientific studies. A lot of it is developed by word of mouth as people gain experience with a method. This leads to differences in opinion about how things should be done, and there is not always a "right" way to do things.

The information in this chapter will lay down some basic principles to guide you to your own success and help you select the best method for your own situation. The various composting methods are discussed in the following chapters.

Location

Where should you compost? The first choice is to decide if you are doing this indoors or outside.

If you are doing this indoors, you are limited by the number of composting techniques that are suitable. Most methods take too much space or produce too much of a mess for indoors. The actual location inside is not critical, provided you can provide the right temperature.

Outdoor composting can be done anywhere, but some locations are better than others. Take the following into consideration.

- Water: Placing it near a water hose will make watering much easier.
- Close to the vegetable garden: A lot of the garden waste comes from the vegetable garden, and that is where most people want to use the finished compost. Placing it there is less work.
- Sunshine: In cold climates you want as much warmth as possible to speed up the process, so site it in full sun. In warm climates too much heat is a problem as it dries out the compost too fast, making shade a better option.
- Flat area: Makes it easier to work and build the system.
- Neighbors: Composting does smell a bit, and putting it right next to your neighbor's dining area might not be a good idea.
- Drainage: Don't place it in a low spot that has standing water.
- Truck access: If you plan on hauling a lot of materials in from other places, make sure your vehicle can drive right to the drop-off point.

Before you start building compost bins, check your local municipal laws. They might restrict what and where you can do it, although most municipalities are now encouraging composting.

Storing Input Ingredients

One of the biggest problems for many home gardeners is storing the starting materials until you are ready to compost. Hot compost piles work best if you build the whole pile at one time, but input ingredients are available at different times of the year.

Grass clippings and kitchen scraps are available all summer long and should be mixed with a higher carbon material. But high carbon material is most abundant in fall when leaves are falling and plants are dying back. To create the right C/N mixture, you need to store some material until you can gather enough of everything to make the pile.

You not only need a place to compost, but now you need a place to store material. That is great if you have the space, but many don't, and a lack of space will limit the composting options available to you.

One option is to use one of the colder methods and not worry so much about the right C/N ratio. Simply add stuff as you get it. I use a small plastic bin for kitchen scraps and keep some fall leaves next to it. When I add kitchen scraps, I also add some leaves.

You can also select a system that allows you to keep adding material on top as you get it, and harvest finished compost from the bottom. To be honest this sounds like a great solution on paper, but it does not work so well in practice because the lower opening is usually too small to get the material out easily.

Another option is to select input ingredients that are all available at the same time. Most gardens have a lot of low nitrogen material in fall. So you can order a load of fresh manure at that time to make a big pile and you are done for the year. Keep your kitchen scraps in a different system (bokashi or vermicomposting), or just add them to the pile. A few handfuls here and there won't make much of a difference.

Air

Most of the composting procedures use an aerobic process, which means that they need plenty of air or, more precisely, oxygen. If they don't get enough oxygen, they become anaerobic and a different set of microbes takes over, slowing the process down by as much as 90%. Anaerobic microbes produce organic acids, amines (derivatives of ammonia), and hydrogen sulfide, all of which makes the pile smell.

You will know that you don't have enough oxygen by using your senses. If the pile smells rotten, you need more air. If it does not heat up and you have a good C/N ratio, the problem is a lack of oxygen.

Air migrates into a pile on its own to a distance of about 18 inches. A pile that is wider than this does not get enough air without some help.

You can add more oxygen by turning the pile or you can use a hand tool called a winged compost aerator. It's a pole with some metal tabs at the bottom. The tabs point down as you insert the tool and bend up when you withdraw it. The tabs then mix the ingredients as you pull it up and down. If the pile is made correctly, you should not need to add oxygen.

Some people also insert vertical or horizontal venting systems to allow more air to reach the center of the pile. Vertical vents can be added with old corn stalks, perforated pipe, or cylinders of wire mesh. The perforated pipe also works well for horizontal vents.

You probably don't need any of these things if you layer your material, use material that does not compact too much, and don't make your piles more than four feet square.

Water

The moisture level is critical. Too much moisture (>60%) and the pile becomes anaerobic because there is no room for enough oxygen. If the material is too dry, microbes die off and composting stops. The optimum moisture level is 45–50%.

Add some water as you are building the pile. Add a few inches of material and water. Add some more material and water again. That way you make sure there is enough water in the center of the pile. Keep in mind that fresh green material is about 80% water, so you need to add less. Course material dries faster and needs more water added than fine material.

How do you tell if you are in that sweet spot of 45–50%? Give the material a squeeze. It should feel like a well wrung-out sponge. If you can squeeze out a drop or two, it is perfect.

Depending on your location, rain may or may not be a problem. In wetter areas, it is a good idea to keep rain out of the pile so that it does not get too wet. Many people will put a roof over the pile or simply cover it with a tarp. If you use a tarp, it is important that you leave an air space between the tarp and the top of the pile to allow oxygen to get in. A few thick branches or a couple of 2 × 4 pieces of wood will provide enough of an air gap.

If the pile gets too wet, take it apart and let it dry out. Then pile things together again.

Turning

Compost is created by mixing a number of ingredients together and then waiting for the magic to happen. Alternate layers of different material and make each layer no thicker than two to six inches. You can then leave the pile until it is time to turn it. Alternatively, mix everything up at the beginning so high nitrogen material is close to low nitrogen material and then make the pile.

Once a hot pile reaches 145°F (63°C), it is overheating, and it is a good idea to cool it down by turning it. This moves the less composted material near the edge of the pile into the center, slowing down microbial activity.

A hot pile will also cool down on its own as it gets past the composting peak. If the temperature drops to 100°F (38°C), turning it will start a new composting cycle and it will start heating up again.

How often should you turn it? That depends on the method you are using and the material in the pile. Cold piles can be turned once a month, or not at all. Warm piles can be turned once a week. Hot piles should be turned when they reach their maximum temperature.

Provided you don't overdo it, turning a pile will speed up the overall process, especially for the material in the outer edges. If you are not in a hurry, you don't have to turn it. A pile of leaves that is never turned will still make compost.

It is important to turn a pile that starts to stink because it is going anaerobic. Turning will increase the oxygen in the pile.

There are a number of different ways to turn your pile, and some composting methods are easier to turn than others. The attraction of tumblers is that they are designed to make turning easier. You just go out every few days and give them a turn.

The three-compartment hot composting bin system was designed to make turning easier. You generally use one bin for new compost material and a second for the current compost pile. When you are ready to turn your pile, you move it from the second one to the third one, mixing as you go. That way the stuff at the bottom ends up on top and outer material can be moved to the center.

The same process can be used for piles that are just sitting on the ground. Just create a new pile as you turn it.

The best tool for turning compost is a fork. A shovel has trouble getting through the course material, but the tines of a fork are perfect.

Using a Compost Thermometer

With all this talk about temperature, you might be wondering how do you measure the temperature? There are special compost thermometers that look like a meat thermometer with an extra-long probe that you stick into the pile.

How necessary is this? Consider it a fun thing to do, but you don't have to measure temperatures. I have been composting for a very long time and have never measured the temperature.

Another option is to carefully stick your hand in the pile. If it is very hot, it is too hot and it's time to turn it. If it was hot a few days ago and now it is cooler, it is time to turn it again. Don't become a slave to the compost pile.

Shred Input Material

Microbes are extremely small compared to the material being composted, and they have trouble getting through the outer epidermis of plant material. It is a good idea to shred things up to make it easier for them. Smaller bits of material will compost faster.

A garden-sized chipper works well for compostable material, but I found they are almost useless for woody branches.

You can also lay the material on the lawn and run over it with a lawn mower or put the material in a garbage can and use a string trimmer to chop things smaller. Both of these work, but if your tools are gas-powered, I have trouble believing this is an eco-friendly option. Even battery-powered devices need electricity which has to be made somewhere. The alternative is to compost a bit slower and be nicer to the environment.

The Curing Stage

Curing starts when turning no longer heats up the pile. The material has reached a stage where much of the readily compostable material is partially decomposed. There is not enough food to support a large population of microbes, and even though they are still active, there is not enough activity to generate heat.

In a hot compost pile, the curing stage usually lasts three to four weeks, but it is a good idea to leave it for a couple of months after it looks finished. It is a very important and often neglected part of the composting process, and it is more important for piles that are not made and managed perfectly.

Immature compost can contain high levels of organic acid, a high C/N ratio, or other characteristics which can damage plants. Degradation continues at a slower rate during curing, which results in a more chemically stable end product.

A long curing time is less important when the compost is used in the garden and more critical when used in pots and containers, especially for seedlings.

Curing also has a downside because it allows more nutrients to leach out of the pile. If you are applying it as mulch, you don't need to cure it. If you are digging it in just before planting, curing is more important.

Using Activators

The term "compost activator" is used in two different ways in the composting community. One use of the term is to describe high nitrogen input ingredients, things like grass clippings and urine. These items are good for a pile provided that you don't overdo them. I am excluding these from the following discussion.

The other use of the term is for a range of products that are being sold to speed up and improve composting. They are known by a range of names, including: compost starters, compost activators, compost boosters, and compost accelerators. Some of these products are just microbes and go by names including: microbes for soil, soil probiotics, soil enzymes, and beneficial microbes.

These products contain fertilizer and/or microbes.

The material used for composting is covered in microbes. You do not need to add more. If you really don't believe me, throw in a couple handfuls of soil from the garden. It costs nothing and has as many microbes as any purchased bottle.

What about the fertilizer? It is unlikely that you need phosphate or potassium since plant material already contains these. Besides, potassium is very soluble and will just be washed out of the pile. If you need these nutrients for your soil, add fertilizer directly to the soil.

Nitrogen is a different story. If you have too much carbon, you do need more nitrogen. If that is the case, use a fertilizer where the first number in the NPK ratio is high. One of the best choices is urea, which is 46-0-0. If you want to go organic, try some blood meal at 12-0-0.

Some products claim to contain an "energy source." I guess they think you need this extra energy to wake up the microbes so they will do their job. What these manufacturers fail to realize is that almost everything in the compost pile is an "energy source" for microbes.

Some people talk about adding extra sugar, molasses, or milk for the same reason. Feel free to add such food if it can no longer be consumed, but don't add it as a special starter for microbes.

None of these special compost activators are required.

Nitrogen Sources

Most home gardeners have lots of high carbon material but not enough high nitrogen material. Manure is a great choice for extra nitrogen, but if you don't want to use it, here are some other good choices.

- Urea (46-0-0)
- Urine
- Blood meal (12-0-0)
- High nitrogen fertilizer
- Cotton meal (7-2-1)—cottonseed can contain herbicides.
- Fish meal (8-6-0)

- Alfalfa meal (11-1-2)
- Feather meal (13-0-0)—degrades more slowly than blood meal.

Wintertime

Composting depends on active microbes, and microbes are only active in warmer weather. In cold climates, not much happens in winter, but microbes are able to survive winter and will become active again in spring as things warm up.

There are a few things you can do to help the process. Compost bins with an open front can be faced south. This collects more heat in fall and spring, at least on sunny days.

You can also wrap bins and other closed systems with insulating material like bubble pack plastic, Styrofoam sheets, leaves, bales of straw, or burlap. These trap heat inside, effectively extending the season.

Winters in zones 6 and colder are just too cold, and no matter what you do, composting stops in winter.

Two Is Better than One

A lot of people will get one composting bin or tumbler, but this has limitations. Let's say that you fill the device to start your first batch. What do you do with new material and kitchen scraps as they become available? If you keep adding material to the first composter, you will never have finished compost.

The solution is simple, you need a second composter.

Give some thought to this problem before you make a decision on which composter to buy. You might not want to pay for two expensive units.

Speeding Up the Process

It is not necessary to compost quickly, but it does allow you to process more material over time. So what can you do to speed up the process?

- Use a C/N ratio of 30:1
- Add a bit of extra nitrogen
- Keep the pile warmer

- Maintain moisture levels
- Chop up the input ingredients
- Use a larger pile
- Cover the pile until it gets warm
- Use easy-to-digest material

Keep Animals Out

There seems to be a real concern about keeping both pets and wild animals out. I don't really understand why. If they spend some time in the pile, they might just add more high nitrogen material for free. Having them dig in the pile is not a problem either—it needs to be turned anyway.

I understand that animals might make a mess spreading material around, but once they find what they are looking for—maybe a hiding rat—they will leave.

Keep meat, fish, and dairy products out of the pile and you are less likely to have an animal problem. And if you do, ask yourself if it is really a problem.

Keep Out Insects

This is another concern I don't worry about.

Flies may lay eggs that form maggots, but beetles will eat them. If you add a lot of fruit, you might get fruit flies—so what? Snails and slugs may take up residence, but that is a good thing. It keeps them off your prized lettuce. Ants won't build a nest in a pile that has the correct moisture level, but they might do so in a dry pile. That's free aeration.

As the compost reaches a mature stage, the only remaining insects will be ones that live in the soil anyway. Insects are part of nature and gardening. A few in your compost pile is not a problem.

Troubleshooting

Composting is fairly simple, but things can go wrong. Here is a list of the most common problems along with their solution. With some practice, you won't have any of these issues.

Symptom	Problem	Solution
Ammonia smell	Nitrogen level is too high.	Add more high carbon material or flush some of the nitrogen out with water.
Rotten egg smell	The pile has gone anaerobic and needs more oxygen.	Turn the pile or manually aerate it using a winged compost aerator.
	Moisture level is too high.	Spread the material out so it can dry. Then rebuild the pile.
Not heating up	Air temperature is too low.	Cover with a blanket or other insulating material until heat can start to build up.
	C/N ratio is too high.	Add more nitrogen.
	Too dry.	Microbes need moisture to grow. Increase the moisture level.
	Pile is too small.	Small piles don't heat well. Accept the lower temperatures.
	It's done.	Wasn't that simple?
Cooling off	Active microbes are running out of food.	Turn the pile.
	Composting is nearing the end of the process.	Do nothing—it is working fine.

Selecting the Right Composting System

Up to this point, the book has focused mostly on hot composting, but much of the discussion also applies to cold and medium composting systems.

Now that you understand the process and the requirements for it, it's time to look at actual composting systems so that you can select the best one for you. The next two chapters focus on outdoor systems that use hot, medium, or cold methods. The following three chapters are on indoor composting systems: vermicomposting, bokashi, and eco-enzyme.

I suggest you read through each section because several of these systems can be combined, and even if you think you already know which system you want to use, it is a good idea to understand them all.

There are many variations to each described system. Once you understand the basics of a system, you will be able to change it, improve it, and adapt it to your specific situation. There are many innovative gardeners online who will show you new variations to consider, and in a couple of years, you might be one of those inventors.

6

Piles, Bins, and Tumblers

This chapter will look at a variety of outdoor composting systems. Some are really simple, others require some construction work, and some can be purchased as ready-to-go systems. There is a wide range of possibilities.

Most gardeners will choose one of the systems in this chapter. That does not make them the best, but it does mean they work for many people. One reason these are popular is because they are based on the time-tested hot composting process that has been used for hundreds of years.

Simple Piles

By simple I do really mean simple. Pick a spot in your garden. Gather your input ingredients. Pile them up and you're done. A bigger pile gets hotter, and a smaller pile takes longer. If you want it to get hot, aim for a pile that is five feet wide and three feet tall, but any size of pile will compost.

Simple Piles

Pros	Cons
Simple, inexpensive, and easy to start.	In windy locations, stuff may blow around.
Easy to turn.	Not the nicest-looking system.
Can make hot compost.	

A simple compost pile.

You don't have to turn it, but if you do, you simply make a new pile beside the first one.

I told you it was simple.

Bins and Boxes

This is your traditional system. The terms bins and boxes refer to the same thing, and I'll call them bins. You can have a one-bin, two-bin or three-bin system, and you can have even more if you need them. Each bin is about a four-foot cube. This is not too large to fill and yet is large enough to make hot compost.

Each bin has three or four walls. Some people like to use four walls so that the front is covered. Others like to keep the front open for easy access. I always thought the front needed a wall to keep stuff inside the bin, but in my last set of bins, I left the front open and it works almost as well.

If you place the second and third bin right next to the first one, you have fewer walls to build, since two bins will share one wall between them.

The walls can be fancy or quite plain. They can be permanent or removable. They can be solid or have openings to allow better air flow, which I prefer. Wood, concrete, and straw bales are common materials for building the walls.

Common three-bin system.

Bins made from wood are the most popular, and if you Google images for wood compost bins, you will find a lot of designs. Pick one that you like and suits your carpentry skills.

If you don't want to invest too much money or effort, consider using used skids. This is the method I use now.

Find yourself some used skids. If you look hard enough, you will find free ones. You need three for the first bin and two more for each additional bin.

Skids normally have three larger boards running one way with numerous thin ones running perpendicular to them. Set one up as the back of the bin so it is sitting on the larger boards. Place another

Bins and Boxes

Pros	Cons
Large enough to make hot compost.	Requires more space than other systems.
Can look quite neat, if you take time to build a nice bin—skids don't look as nice.	Extra effort to make them.
A three-bin system provides room to make compost and collect material for the next batch.	

one at right angles to the first and join the corner. You can screw them together, nail them together, or just use wire to tie them tight.

I cut some extra scrap wood, placed it across the corner, and screwed this into both skids to make a solid corner. The lower section was just wired together.

Add the rest of the skids in the same way. You now have a multi-bin composting system with open fronts. You can add fronts if you want but design them so they are easy to remove, making it much easier to remove finished compost.

Wire Cage

This is a simple way to make a good hot composting system. Take some wire fencing material and form it into a three-to-four-foot-wide cylinder that is about four feet tall. Tie the open end together

Credit: Denise Krebs, https://www.flickr.com/photos/mrsdkrebs/5812458475

Wire cage holds compost in place.

with some wire or strong string. It does work better with some vertical posts sunk in the ground, to hold the wire in place. Fill the enclosure and you have a compost bin.

When it is time to turn the pile, remove the wire and set it up next to the first pile. Then transfer the first pile into the new wire bin.

You can use chicken wire for this, but that is a bit flimsy. A thicker wire mesh material is more expensive but works better. Cattle or swine wire fencing makes a really strong bin. A wooden or plastic snow fence can also be used.

Wire Cage

Pros	Cons
Easy to make and easy to move to a new location.	Needs to be taken apart to turn the pile.
Large openings allow a lot of air movement.	May dry out faster than other systems.
An added top will keep larger animals out.	
Large enough for hot composting.	

Plastic Garden Composter

This group of composters are also called stationary composters and compost bins. They are made from plastic and are two to three feet tall. They usually have a removable lid to make it easy to add material in the top and a door at the bottom for removing finished compost. The sides have vent holes to allow air to get in. There are many commercial versions available.

These units are designed to sit in one location. You can fill them all at one time and do more traditional composting, but many people just add new kitchen scraps on top every few days and take out finished compost from the bottom.

These bins look nice and are easy to use, but they have some limitations. They are too small to heat up, and most do not make it easy to turn the compost. You can lift the composter off the pile, place it in a new location, and then turn the old material into it again.

Credit: Snowmanradio, https://commons.wikimedia.org/wiki/File:Compost_bin_16l07.JPG

Plastic garden composter.

I use one outside the back door of the garage for kitchen scraps. I have old leaves nearby and throw some in with every addition from the kitchen. I find the door at the bottom too small for removing finished compost. Instead I wait a year and move the whole pile to the garden all at once.

Plastic Garden Composter

Pros	Cons
No construction needed and readily available. They can be expensive, so buy a used one.	Too small to get hot.
Looks nice in the garden.	Makes compost very slowly.
No rodent problems in a closed system.	Most of these systems do not have enough air vents, so be careful about getting it too wet.

I also found mice living in mine, maybe because of the leaves, so I put down some wire mesh on the ground under it to keep them out.

If you don't want to buy a fancy composter, take a large plastic garbage can that has a good fitting lid and drill some air vents in the can. Fill it with material and let it compost. This will work just as well as a commercial product, but may not look quite as nice.

Tumblers and Rotating Drums (Barrels)

The big concern about plastic garden composters is that it is difficult to turn the material. To solve that problem, people have invented various composters that are easy to rotate. A very simple device is a drum or garbage can with a lid. Poke some holes in the side for ventilation, fill it with scraps, and lay it on its side. Then every few days roll it around the yard to mix the material inside.

You can also get a drum that fits on a base that has small wheels to make it easier to rotate the drum. These usually have hatches on the side for filling and emptying.

Rotating these large drums can still be difficult once they are full of wet organic matter. To solve this problem, the drum can be raised up above the ground and fitted with a handle to make rotating

Credit: Tom, https://www.flickr.com/photos/amayzing/3100327938

Twin compost tumbler made by Mantis.

easier. These are available commercially, and there are numerous DIY designs on line.

Some of the commercial products claim that the "tumbling" action produces compost in two weeks. That's nonsense. These devices do make compost and they are easier to turn, but turning them too frequently slows down the rate of decomposition because it does not give the material time to heat up. These units tend to be slow.

The ability to turn and mix the compostable material is a benefit over stationary devices. Just don't turn them too much.

Tumblers and Rotating Drums

Pros	Cons
Easier to turn.	Too small to get hot.
Easier to empty if they have a door or hatch.	Makes compost slowly.
No rodent problems in a closed system.	Most of these systems do not have enough air vents, so be careful about getting it too wet.

Food Waste Digester

Natural composting takes place on top of the soil and in the soil. Why not mimic this process? That is exactly what food waste digesters do.

These digesters consist of a can that is sunk in the ground. It has a lid above ground for adding compostable material, and the portion below ground has numerous holes to provide access for natural soil organisms.

You can make your own system or buy a commercial product. The Green Cone is a popular commercial product. It has a plastic bucket (the can) that sits below ground and a plastic cone-shaped device that sits above the can. Waste material is added into the top of the cone. According to the manufacturer, up to ten pounds (four kilograms) a week can be digested in this system. I find that hard to believe since digestion does not happen that fast.

The Green Cone composts below ground.

The Green Cone also promotes the fact that you can digest meat and bones because it is a closed system that keeps animals out.

You can make your own system using a metal or plastic garbage can with a good-fitting lid. Drill 30 one-quarter-inch (six mm) holes in the bottom of the can. Now drill another 40 holes in the side of the can, but only in the lower half. Dig a hole in your garden that is a bit wider than the can, and deep enough to cover half of the can so that all the holes in the can will be below grade. Place the can in the hole and fill soil around it. You are ready to add the food waste.

As you add the compostable material, include a handful of soil and keep the can covered with the lid.

One problem with this system is that turning is next to impossible. Manufacturers claim you don't need to turn the material. You can dig down in the can to turn things or to remove finished compost from the bottom, but it is probably easier to stop adding material after six months and start a new can. When the first can is finished composting, it can be emptied, ready for use when the second one is full.

Another option is to lift the can once it is mostly full and dump the contents into the hole. Cover with soil and place the can in a new location.

In cold climates, one advantage to composting underground, where it is warmer in winter, is that composting happens over a longer period of time.

Food Waste Digesters

Pros	Cons
Very easy to use.	Too small for hot composting.
May work better in fall and winter than other systems because it is underground.	Difficult to turn or remove finished compost.

Windrow Composting

Windrow composting is done on larger properties, farms, and in some municipalities. They process too much material for simple compost bins, so they pile it up in long rows and use machinery to turn the pile.

This is hot composting that is highly controlled to make a good quality of compost, in a short period of time.

Windrow Composting

Pros	Cons
Produces large amounts of compost.	Needs more space and more compostable material.
Easy to manage a hot process.	

Sheet Composting

A number of terms are used interchangeably for this method, including sheet composting, sheet mulching, lasagna composting, and lasagna mulching. These are used to describe two different gardening techniques, and the terms are routinely used incorrectly.

Lasagna gardening is a process where you make a pile of organic material by layering each input ingredient, hence the name "lasagna." Over time these materials decompose and can then be mixed into the soil. It is really a compost pile that is made on top of

Sheet mulching with newspaper and cardboard.

the soil you want to improve. It is usually no more than six or twelve inches high, and it may or may not include paper or cardboard.

The technique I just described has been used for hundreds of years and was originally called sheet composting. In more recent times, gardeners have started calling it lasagna gardening or lasagna composting, and these are now more popular terms.

The term sheet mulching is now used for a variation of lasagna gardening that is used for killing a lawn and making a new garden bed. A layer of cardboard or newspaper is laid on top of the grass or weeds. Water is used to wet it down and prevent the wind from blowing it away. Some heavier material, like soil, compost, or mulch,

Sheet Composting

Pros	Cons
A very organic way to kill grass or make a new bed.	It is a slow process.
	Soil is unavailable for planting until things compost.

is then layered on top of the paper. Once the paper is decomposed, the bed can be used for planting.

The paper or cardboard kills the grass and most of the weeds under it in about four weeks, depending on temperature. I prefer to use newspaper because it decomposes quicker, but even this can take a couple of months in summer to decompose properly. Cardboard also works, but it will take a year to decompose in most climates.

Some people start planting right away, by making slits in the paper. This does work, but it has a serious downside. Grass will grow through the slits and grow right in the center of perennials, where it is impossible to remove. For permanent landscape borders, it is best not to plant until the grass is dead. If you are planting annuals or vegetables, this is not as big a problem because you can remove any remaining grass the following spring when you make the new garden.

The process works very well for killing grass and many weeds, so some gardeners use the paper as annual mulch, but that is really not a good idea. Firstly, paper adds almost no nutrients to soil. Secondly, it does not allow water to reach plant roots until it decomposes. Use it the first year to make a new bed, but don't use it as mulch in future years.

Compost Myth: Sheet Mulching Harms Soil

It is believed by some that sheet mulching reduces the air exchange with soil, which in turn harms soil biology. As far as I am aware, there has only been one study[1] that looked at this issue, and it concluded that cardboard did not change the rate at which air is exchanged with soil. This work was done in pots, so the results can't easily be extrapolated to field conditions, but there is no science that supports the idea that the paper harms soil or soil biology.

The technique has been used by many thousands of gardeners, and I have never seen a report that suggests plants do not grow well using this technique to prepare new beds. It is not a technique that should be used annually around permanent plants.

7

Easy Composting

So far in this book, I have described more formal ways to compost, and all of them are great, but they tend to be more work than some of us want to do. This chapter will focus on other techniques that are perhaps more suited to the lazy (or is that smart?) gardener.

Cut and Drop Method

About 15 years ago, I moved to a six-acre property and started developing a large botanical garden. One of the first tasks was to build a traditional two-bin compost system. It worked great, but I found that it was a lot of work moving material from the garden to the pile, and then in spring I needed to move it back to the garden again. To save time, I started using the cut and drop method, and I love it. I now use this technique for 95% of my yard waste.

The method is simplicity itself. Wherever you are in the garden, leave the plant waste where you stand.

I do most garden cleanup in spring. My hosta leaves are already on the ground from winter. They stay there. The old hosta flower stems might still be vertical, so I take some hedge trimmers and cut them off, about 6 inches at a time. The cuttings are left where they fall.

Any other dead foliage is treated in a similar way. Just cut it and drop it.

An exception to the rule is some tall ornamental grasses that make thick stems. I can't be bothered cutting them all up into small pieces, so I cut them off at ground level and haul them to a compost pile. I also treat rose bushes differently. I cut those back and put the cuttings in our municipal green bin. I hate thorns.

If I am deadheading or cleaning up plants in summer, I just cut and drop. If the plant is near the front of the bed, I fling the cut piece behind a larger plant at the back to keep things looking neat.

In fall, all leaves are left where they drop, provided it is in a flower or shrub bed. On the grass, I either mow them into small bits and leave them or rack them into a nearby flower bed that needs some extra organic matter. I don't rake them very far.

Most weeds are either annuals or perennials that don't spread by runners. All of these are just pulled and dropped into the garden bed. If I have concerns about them rooting, I will leave them on the garden path for a couple of days to dry out in the sun, and then I throw them into the flower bed.

There are exceptions. Invasive weeds like bindweed, quack grass, and Canadian thistle are not returned to the garden—they are just too nasty! They either go out with the garbage or they are dropped on the lawn to die.

What about lawn grass that invades flower beds? I pull it out and toss it onto the lawn. It either dies or roots—either is OK with me.

Nature has been composting organic matter that sits on the surface of the soil for thousands of years. Why not let her take care of the composting for you?

Thank you, Mother Nature.

Cut and Drop Method

Pros	Cons
Very easy and no worry about the C/N ratio.	Can look a bit messy in very early spring, but plant growth soon hides it all.
Nothing to buy.	

Leaf Mold

Leaf mold is made by piling up leaves in the fall, and leaving them alone. Since leaves have a high C/N ratio (60:1), they decompose slowly and the pile stays cold. Most of the decomposition is done by fungi rather than bacteria.

The fungi tend to come from underground and grow their mycelium up into the leaf pile where it decomposes the high carbon material. The end result is a combination of broken up leaves and white mycelium which makes great compost. It is best to use it in the garden as mulch.

The process can take two years in colder climates, but shredding the leaves, keeping them moist, and adding nitrogen will speed up the process.

Fall leaves are fairly low in nutrients since the tree removes some nutrients before the leaves drop.

Piles of leaves can also be messy, especially in windy locations. To help with this problem, the pile can be contained in a wire cage, as described in the previous chapter.

Leaf Mold

Pros	Cons
Easy to make.	Takes a long time.
Good soil amendment.	Can be messy in a windy location.

Pit and Trench Composting

Nature can compost material all on its own by leaving it on the ground, but many people don't like to see the uncomposted material. A simple solution is to bury it. A pit is a round hole, and a trench is a longer rectangular hole, but they are the same technique.

Find a suitable spot in the garden and dig a hole big enough to hold the material you have. Fill the hole with compostable material and cover with soil. If the material is high in carbon, you can add a bit of nitrogen (urea, blood meal, etc.). Then forget about it.

The microbes and earthworms in the soil will soon find the material. Over time it decomposes and is absorbed into the soil. Nutrients will migrate out of the hole into the surrounding soil where plants will find them. The plants will also grow roots right into the compost once it is finished enough. This technique is not really suitable for producing compost that you want to use somewhere else.

In my very first vegetable garden, I used raised beds without side walls. The pathways were used as a trench, and I slowly filled them with yard waste. In the following year, I would empty the pathway and place the finished compost on the beds, ready to start all over again.

If you plan to trench the material near other plants, especially vegetables, it is a good idea to dig the trench early in the season before plant roots grow into the space. You don't want to dig a trench near a tomato plant mid-season and harm the roots.

The system works well for smaller quantities of material like your kitchen scraps. It is more difficult to deal with large quantities of material this way.

Pit and Trench Composting

Pros	Cons
Easy system to use.	Need to find places to dig up in your garden.
Works well for kitchen scraps and small amounts of yard waste.	Less suitable for large quantities of organic material.

Keyhole Composting

Keyhole gardening is both a composting method and a gardening method. I am going to describe the more traditional design of this method so you understand the whole concept, and then I'll describe some variations that are easier to use.

The whole thing starts by building a special keyhole garden which is a circular raised bed that contains a center composting area. One side of the circle has a pie-slice cut out of it so that you can reach the center. I guess the term "keyhole" comes from this shape, but I don't really see a keyhole?

Credit: Julia Gregory: https://www.flickr.com/photos/yarnmaven/7006331666

Keyhole garden made with cinder blocks.

If you Google keyhole gardens, you will find many different designs for such a garden. The raised bed can be any height, and the wall can be made out of any material, including rocks, bricks, or even wood. It does not even have to be a circle. It could be a square with one side indented to allow access. The garden can also be domed with the center higher than the walls.

The center composting area is some type of cylinder or cage that can hold material. I have seen people use wire mesh, steel drums, and plastic buckets. The requirement here is that the bottom is left open so the compost sits on soil, and the sides should have lots of holes to allow soil organisms access to the organic matter.

This design is essentially a garden built around a composting cage. New organic matter is added into the top of the cage as you collect it. It slowly moves down as older material decomposes. Nutrients leach out the side into the garden, and macroorganisms can move from the soil into the compost for a snack. The cage itself is either never emptied or only emptied every few years.

This composting process is quite simple, but making the garden is more work. There are ways to simplify the system and still retain most of its benefits. The raised bed can be only a few inches high and made with a single row of bricks. The center compost pit would need to be lowered so composting happens below grade.

Keyhole gardening will also happen even if the garden bed is not raised, or raised a few inches with no wall around it. The main point is to have a garden all the way around the composting pit so that the whole garden benefits from leached nutrients.

This is an interesting idea, but I think it has some fundamental flaws. The claim is that nutrients flow from inside the cage into the surrounding soil. Although this will happen to some extent, most nutrients move down in soil, not sideways. Plant roots live mostly in the top six inches of soil. That means plants won't get most of the nutrients.

Reports that I have seen show plants doing equally well over the whole garden, but you would expect the ones near the center to do much better if they were getting nutrients from the compost, so I suspect the plants are getting their nutrients mostly from the original soil and not from the compost.

Raised beds dry out faster and therefore require more watering.

If you would like to build one of these gardens, consider building it very close to grade and don't raise it up more than six inches. The compost cage in the center should be placed at grade so most of the material is above the garden, and it should have a solid bottom. These changes will reduce watering requirements and increase the nutrients that make it to plants.

Keyhole Composting

Pros	Cons
Easy composting.	Nutrients will leach deep into soil away from plant roots.
	Requires more watering.
	Structure needs to be built

Electric Composters (Food Digesters)

I first saw an electronic composter for sale about six years ago, and now there are at least a dozen models on the market. These devices grind kitchen scraps and heat them up to drive off the water, producing a compact form of the kitchen scraps.

Some systems require the addition of other ingredients like coir (crushed coconut husk), and some even promote the idea of adding microbes. The newer systems don't require either of these.

These systems make the following claims.

- They compost your food scraps.
- They turn food into fertilizer.
- They reduce the volume of food scraps.

These devices do not compost anything. They grind and dry your food. Any such product that claims to be composting is misleading its customers. Their process takes a few hours, and composting takes weeks and months.

Many of the products claim to "grind" the food as it is being heated. The blades in the unit are not sharp enough, nor is the space

A commercial electric composter.

The so-called fertilizer from an electronic composter. Pieces of food are clearly visible and identifiable.

between the blades and the fixed bar small enough to grind food. They rotate once per minute, and at best this can be called agitation. This is also very clear when you look at the results. The material is not finely ground and contains a lot of stringy material.

It is unclear what these devices produce. I have talked to several manufacturers, and none have given me a clear answer. Instead they use a lot of marketing gibberish. As far as I can tell, these devices produce dehydrated kitchen scraps. It is not fertilizer, it's certainly not compost, and it is only partially ground up.

I was able to get some analytical data from one company that seems to be more ethical than most, the Vitamix FoodCycler. They show the NPK of the ground food to be 2.9-0.2-0.8. The 2.9 is the total nitrogen, and the available nitrogen is 0.005%, confirming that composting has not yet started.

Commercial food digesters are expensive to buy and require electricity to run. Most units have some type of odor filters that need to be replaced every six months, and they are not cheap.

The products recommend adding the dehydrated kitchen waste directly to your potted plants, but I would not do that. Since it is not decomposed yet, it will draw nitrogen away from plants to start that process. When I tried mixing it with soil, the material rehydrated and started to grow a lot of mold.

It would be much better to add this material to your composting system or use it as mulch.

Electronic Composters

Pros	Cons
Can be used indoors.	Handles very small amounts of material, not suitable for quantities of yard waste.
Finished product can be stored and used on indoor plants.	Very costly for the quantity they are able to process.
It is an option for indoor processing of kitchen scraps in cold weather.	

8

Vermicomposting

Vermicomposting uses a bin, organic material for food, and worms. The worms eat the organic material and produce worm castings, i.e., worm poop. The worm castings are then used as fertilizer for potted plants or the garden.

Unlike other composting methods, this one is as much about keeping worms as it is about making compost. You have to care for your new pets, which are more work than a static compost pile in the garden. On the flip side, this is a great method if you want to produce compost indoors.

Although it is called composting, it is not really a form of composting. It should be called vermidigestion, not vermicomposting. As food passes through the worm, it is ground up and partially digested. There is almost no composting taking place.

The material that exits the worm is finely ground organic matter full of microbes, especially bacteria. What happens next? It starts to decompose, and after a few months, it has turned into compost.

Vermicomposting is really two systems in one: a worm poop factory and a compost pile.

Since the process usually takes place in a container, it is suitable for small amounts of organic matter, especially kitchen scraps, and that is how it is usually used by home gardeners. Most gardening discussions on vermicomposting do not recommend the method for yard waste, but agriculture looks at vermicomposting differently and processes large piles of manure and yard/farm waste using worms.

The process can be done outside in warm climates, but in temperate climates, it works best indoors since the worms that are used can't survive a cold winter outdoors.

Understanding Worms

Worms are cold-blooded invertebrates (an animal with no backbone). They are quite different looking than humans, but there are also some similarities. They breathe in oxygen and exhale CO_2 just like us. They have a mouth and a digestive system although it is much simpler than ours. They eat many of the same things we do, making them ideal for eating our kitchen scraps. In nature they eat many types of plant material.

They have no eyes, but they can sense light and are photophobic—they prefer darkness. They have no ears, but they can sense vibrations which stress them out. They know when you are working on their home.

Worms do not have lungs, so they breathe through their skin. For this to work, the skin has to be moist all of the time, so they create a mucus to keep from drying out. They also can't be too wet or they'll drown.

Worm sex is very interesting. They are hermaphrodites, meaning each worm is both male and female. Each one produces both eggs and sperm, but they can't mate with themselves. Two worms need to cuddle so that they can exchange sperm. After sex each partner produces a cocoon full of eggs and then adds the collected sperm.

Baby worms hatch in about two weeks, if conditions are right. If the conditions are not right, hatching is delayed until it is right. In six to eight weeks, the worms are mature and ready to start their own family.

You might have noticed that worms have a bulge near one end. This is the clitellum, which plays a role during sex and develops once a worm is mature.

The cocoons are quite small, about the size of a small rice grain. They are initially a light yellow and darken to a light brown.

Digestive Process

Worms eat soil and organic matter, which travels down a long digestive tube that consists of several key sections. First comes the esophagus that adds calcium carbonate as a way for the worm to rid itself of excess calcium. The food then moves on through the crop and into the gizzard. The gizzard uses swallowed grit to mash food into small particles while enzymes are added to help digestion. The material then moves into the intestine which is two thirds of the worm's length and where fluids are added to further digest the food. Similar to our own intestine, it absorbs nutrients that are needed by the worm.

There is one other key ingredient, microbes. The worm controls moisture and pH levels to favor the growth of a microbial population, including bacteria, fungi, and actinomycetes. These microbes play a major role in the digestion of organic matter.

Along with the soil and organic matter, worms also ingest large amounts of microbes. In fact, the microbes are their primary source of food, not the organic matter.

The whole digestive system is not very efficient, and only five to ten percent of the ingested food is absorbed by the worm. The rest is excreted as mucus-coated particles called vermicasts or worm castings. The worm castings contain undigested plant material, nutrients, soil, and a large amount of microbes. The microbial activity of castings is ten to twenty times higher than in soil or other forms of organic matter.

Composting Process

The worm is not really doing much composting. It does some digestion, but its main contribution to the process is that it breaks organic matter into small pieces and mixes it with microbes. A point that is not emphasized enough is the fact that much of the composting process takes place after the castings exit the worm. This external processing is the most significant part of the process.

Vermicomposting is faster than traditional hot composting.

What Is Vermicompost?

There are a number of terms that are routinely interchanged and need to be clarified.

Vermicasts is a fancy term used for worm poop. This is the material that comes out of the worm. Castings is a short form of the word.

Vermicompost is a mixture of material that includes castings, bedding material, organic material left over from uneaten food, nutrients, worms, cocoons, and microbes. It includes everything you find in a mature worm bin. Many people use this term incorrectly to refer to worm castings.

Vermicomposting is the process of using worms to make vermicompost.

Vermiculture is the process of raising worms.

It is important to understand the difference between vermicasts and vermicompost because they are different products and are recovered differently from the worm bin. They are also used slightly differently in the garden.

As I mentioned above, vermicasts are the product of digestion and not composting. The composting process takes place after the castings are made. One of the reasons vermicompost is allowed to sit and age for a while is to allow this composting to continue.

There are also a couple of other terms that are important to understand: vermicompost leachate and vermicompost tea.

When the worm bin is in operation, it might have too much moisture in it, and this vermicompost leachate will drip to the bottom. Some bins are designed to make it easy to drain this off, which should be done every few days or at least with each feeding. The liquid contains organic matter, nutrients, and microbes. The microbes can grow and make the liquid smell since there is not enough oxygen in the liquid to keep it aerobic. The use of this material is described in the section below called Using Vermicompost.

Once the whole process is finished and you have vermicompost, you can use it to make vermicompost tea by adding some to water and letting it steep. This process is exactly the same as making any other type of compost tea and is fully described in Chapter 12.

Compost Myth: Gardeners Harvest Worm Castings

Let's have a closer look at what you harvest from your worm bin. Many gardeners think they are harvesting worm castings, but that is not usually the case. Home-based worm farmers tend to let their bins run for several months, even up to six months before they harvest.

Let's think about this for a minute. On day one, worms make castings. What happens to these castings? They start the composting process, and after six months, they are highly composted. Castings are also made the day before harvest. So what you are harvesting is a mixture of some fresh castings, a lot of partially composted castings, and some finished compost.

It is more correct to call this vermicompost, than vermicasts. Home gardeners don't really harvest worm castings.

Commercial production can operate on a shorter time scale. Castings can be harvested a couple of weeks after the worm produces them. This produces a higher percent of actual worm castings because they have not had enough time to decompose.

Selecting the Right Worms

There are many species of worms, but most vermicomposting is done with a special species called *Eisenia fetida*, which is commonly known as the compost worm, manure worm, redworm, and red wiggler. It is one to four inches long and lives for three to four years. This worm is tough, adapts well to bins and our homes, and breeds very quickly. It is frequently found in manure piles in many parts of the world.

Common earthworms do not work well for vermicomposting because they want to burrow deep in soil. What are needed here are worms that like to stay on the surface. Here are some other less common worms that can be used.

- Red earthworm (*Lumbricus rubellus*): better able to survive cold winters.
- European nightcrawler (*Eisenia hortensis*): is gaining popularity.

- Indian blue worm (*Perionyx excavatus*): popular in tropical climates.
- African nightcrawler (*Eudrilus eugeniae*): popular in subtropical and tropical climates.

Always check the species name before buying worms.

How many worms do you need to buy? You can start a worm farm with any number of worms. Just make sure that you only add enough food for the worms you have.

Another way to look at this is to figure out how many worms you need to manage the organic matter you will produce. One thousand worms weigh about a pound (one-half kilogram), and they eat one-half pound (one-quarter kilogram) of food a day (fresh weight).

Worm Bins

Farming worms has become popular, which has led to many different types of commercial bins and many DIY projects for building them. If you are just starting out, I suggest going with a simple bin. Once you have done it for a while and like it, move up to a more deluxe version. I am also sure you can find a good deal on used vermicomposting equipment.

A worm bin can be made out of almost any material, including wood, metal, plastic, and even Styrofoam. Bins tend to be moist, so wood may not be the best option. The bin needs to be dark and conserve moisture. It should have lots of vent holes for air circulation. The top can be closed, open, or covered with a screen. The purpose of the top is not to keep the worms in but to keep pests out.

Height is also important, but shallow is better than deep. A large shallow bin is much better than a smaller tall one, and a height of 12 to 18 inches is ideal. One pound of worms (1,000) needs about one square foot of space.

A vent at the bottom for draining leachate is a good idea, but many users keep their bins on the dry side and don't need the vent. If things get too wet, it is difficult to get excess liquid out without a drain.

Here is a list of some popular types of bins.

Plastic Tub

This is a good and inexpensive way to give this hobby a try. Take a standard Rubbermaid plastic tub or similar brand. Just make sure it is not clear plastic and not too deep. Drill lots of one-quarter-inch (six mm) holes along the top edge and some larger one-half-inch (13 mm) holes in the lid. The bin is ready to use.

You can make a few improvements to this design.

Instead of drilling holes in the lid, cut out a larger rectangle and cover it with some hardware cloth or window screening. The hardware cloth can be attached to the plastic with screws, rivets, or silicone glue. This opening provides better air circulation and prevents insects, like flies, from entering the bin.

The design as I have described it so far does not allow leachate to drain out. A simple solution is to also drill small holes in the bottom of the bin and set it on something to collect the leachate. This could be another shallow plastic container, or an old cookie sheet. For even better drainage, raise the bin up by placing some scrap pieces of wood under it.

Metal Bins

Metal bins are not as popular, but they are a good idea if you plan to keep your farm outdoors. Metal is much more rodent resistant than plastic or even wood.

Stacked Trays

One of the problems with a worm farm is that at some point you need to separate the worms so you can harvest the compost. There are several ways to do this, but one simple way is to use stacked trays/bins, also called flow-through worm bins.

The worm farm is started with one tray and used as normal. You add the bedding, the worms, and food. When that tray gets full, stop feeding the first tray and add an empty tray on top of it. Put bedding and food in the upper tray. When the worms get hungry enough, they climb up into the second tray to get the food. You can repeat this with several more trays while the material in the first tray ages.

After three or four months, the compost in the bottom tray is finished and it is removed. It is now free of worms and can be emptied. This empty tray now goes on top as the next new tray. This cycle is continued, providing a continual supply of compost, and you never have to separate the worms from the compost.

The way I have described the process is the way the manufacturers suggest using them; however, users have found that worms are not so cooperative. They like moving down, not up, which means they never completely leave the lower bins. If you find this to be the case, reverse the process.

When the first tray is full, place it above a new tray holding new bedding and food. Leave the lid off and put a light above the bins. The worms will now move down to get at the food and because they don't like the light.

This type of system is available as plastic commercial products, commercial wood products, or you can make your own.

VermiHut Plus stacking worm trays.

To make your own, you can build several wooden trays that stack on top of one another. Add some hardware screening to the bottom of each tray.

You can also make this system with several plastic tubs that nest together. Make sure that the holes in the bottom of each bin are large enough to easily allow worms to crawl from one bin to the next. Or you could cut larger holes in the bottom and glue on some quarter-inch hardware screen. Many of these plastic tubs have tapered sides so they sit inside of one another, making a nice stack.

Long Bin

Commercial vermicompost facilities use long benches for production. They start the worms and food at one end of the bench and over time keep moving the new food toward the other end. The worms follow the food and slowly crawl from one end to the other. The compost can be collected once they leave an area. This idea is easily adopted for home use.

Credit: Erik Knutzen, https://www.rootsimple.com/2011/11/my-big-fat-worm-bin/

Long vermicompost bin with two sides
makes it easy to separate the worms.

Build yourself a long wooden bin. It does not have to be very wide or tall. Design it with a vertical separator in the middle of the box. The separator divides the box into two smaller compartments. Drill some one-quarter-inch (six mm) holes in the separator panel so that worms can move from one side to the other.

Start the worm farm in one compartment and operate it until it gets full. Now place food and bedding into the second compartment so the worms migrate from the first one to the second one. After a few days, all of the worms will have left the first compartment and you can harvest the compost.

Worm Food

Worms prefer a balanced diet just like us—fats, carbohydrates, proteins, and minerals. The food that benefits them the most are the microbes living on the material they eat.

They can eat most organic material in the garden as well as most kitchen scraps, both raw and cooked. They can even eat some weird things like paper and coffee grounds. Root vegetables are hard for the worms to eat unless they are first cooked, but they will eat them in time as microbes make them softer.

They can eat eggs, meat, and dairy, but these items are best avoided because they smell as they age and they attract more pests. Small amounts of these are OK. They don't eat seeds from things like peppers, squash, tomatoes, or cucumbers, but if the seeds sprout, the seedlings will be consumed. Don't feed them hot peppers, other spicy food, greasy or oily food, or very sweet things.

Worms prefer a neutral pH, so most people do not feed them very acidic food, but they can be conditioned to it. Bokashi ferment is quite acidic, but it can be fed provided it is added in small amounts until they adapt to it. Once they are conditioned to it, they seem to enjoy it.

It is best to stay away from salty foods. Too much may harm the worms, but equally important is the fact that you don't want high levels of sodium in the final compost because it is quite toxic to plants.

Compost Myth: Don't Feed Citrus to Worms

Most of the advice says that you should not feed citrus like oranges or grapefruits to worms. The first reason is that the low pH will harm the worms and make the bedding too acidic for them. The second reason is that they won't eat the citrus oil in the skins.

None of this is actually true.

Citrus fruit is acidic, but this acidity is quickly neutralized by microbes as they digest the acid molecules. It won't acidify the bedding unless large amounts are added at one time.

The skins are tough and difficult for worms to eat, but molds quickly develop on them and the mold starts the digestion process. In a few days, the rinds become mushy enough for the worms to eat. At that point, they seem to really like them.

If you run low on food, you can give them more bedding material, which they will eat, or cornmeal which is both good for them and inexpensive to buy.

Worms have no teeth, and they have an easier time eating material that is finely chopped. Food that is too big and hard may be avoided for a few days until decomposition softens it up.

Grit is important for their digestive process, so it is a good idea to add some grit, like sandy soil or ground-up eggshells, along with the food.

Bedding Material

Worms need some place to rest, and bedding material provides such a spot. Good bedding material will have these qualities:

- Holds moisture: worms need to stay moist or they suffocate. Most of that moisture is provided by the bedding material.
- Provides lots of air spaces: access to fresh air is critical so the bedding needs to be bulky and should not pack down too much.
- Has a high C/N ratio: a high carbon level ensures that the bedding breaks down slowly and does not heat up. Some heating at

the food source is OK, but worms need a cool place to hang out after a big meal.

• Has a neutral pH: worms like a pH around 7.

Bedding is normally added when a new bin is started. The worms will leave the bedding alone if better food is provided. When the food runs out or they don't like the food, they will eat the bedding until they get more suitable food.

The following make good bedding material: shredded newsprint, shredded cardboard, fall leaves, manure, wood chips, coconut coir, straw, peat moss, seaweed (with salt washed off), and sawdust.

If you are using paper, stay away from glossy paper and paper that contains colored ink. Most black-and-white printing uses food dyes and is perfectly safe for worms. Newspaper is ideal. Simply rip it into two-to-three-inch strips and add it to the bin. Paper from an office shredder also works, but it doesn't hold water as well as newsprint.

Leaves are a great source in fall, but they have a tendency to mat together. They work better if they are chopped into small pieces.

Manure is one of the most popular bedding materials for larger outdoor worm farms. Composted manure is a better option for home use.

Wood chips are not eaten by the worms and can be reused. They can also be mixed in with other material that has a tendency to pack down. They don't hold much moisture until they age.

If you use coir, make sure it was washed by the manufacturer to remove salt. If you are unsure, you can always mix it with water, let it sit, and pour off the water before using it. This should remove any remaining salt.

Peat moss makes good bedding, but it is acidic. Some people add a bit of lime (calcium oxide) to neutralize the acidity, but don't use hydrated lime (aka slaked lime), which is calcium hydroxide.

Dampen the bedding before using it. You don't want it too dry or too wet. It should feel like a damp sponge, and no water should come out of it when you squeeze it.

Start the bin with a three-to-four-inch layer and add more if it gets eaten or if the ratio of bedding to castings gets below 50%.

Caring for Your Pets

It is important to understand that vermicomposting is more about keeping pet worms than operating a compost pile. They need regular attention. Maybe not as much as a cat or dog, but you do need to check up on them once in a while. The good news is that you don't have to get up in the middle of the night to take them for a walk.

What do worms want?

- Food: they like to eat daily, but you can feed a couple of times a week.
- Moisture: the bedding needs to stay moist. Even a short dry period will kill them.
- Oxygen: you should not have to worry about this if you use a good bin and you place it where it gets good air movement.
- Bedding: give them adequate bedding.
- Dark and quiet: place the bin where they don't get strong light or vibrations from appliances.
- Temperature: this depends a bit on the type of worm, but most are happy between 59°F and 77°F (15°C to 25°C). A safe range is 50°F to 85°F (10°C to 29°C).

Start Worms the Right Way

Set up the warm bin before the worms arrive. It should be placed in a spot that is convenient for you to work on them and where you have access to water.

When you get your worms, check them for health. They should be moist and fat looking. Worms that are darker in color or getting thin may be dying. Shine a flashlight on them to see if they move. They should try to hide from the light. If they are not behaving correctly or they don't look healthy, take a video and send it to the supplier right away. Don't assume they will get better.

Add three to four inches of bedding and wet it down. Place the new worms on the bedding, and they should burrow down into it fairly quickly. Newly received worms are stressed and have a tendency to crawl all over the place, so it is a good idea to leave a light on near the bin for a day or two. This will cause them to stay in the bedding and get accustomed to their new home. Once they have settled down, you can turn the light off and they'll stay in the bin.

Don't feed them for a few days. If they're hungry, they will eat the bedding. After that introduce food slowly and add a bit each day. Add more as they consume more. Highly stressed worms may not eat for several days.

Feeding

Some people feed every day, and others feed once a week. Either is fine. The key is to provide enough food so the worms don't go hungry, but not so much that the bin is full of uneaten food.

It is a good idea to add food to a new spot in the bin. That way you can easily monitor the older food to see if it gets eaten.

When you add food, remove an inch or so of bedding, place the food, and cover it with bedding. Covering it places it right where worms like to eat, and it reduces odors and pests.

A thousand worms (one pound) will eat about one-half pound (one-quarter kilogram) of food each day.

When you go on holidays for a short period of time, give them more food and extra bedding. If they run out of food, they will live on the bedding until you get back. For longer holidays add even more bedding.

General Care

Watch your worms and get to know them. Then each time you visit, give them a health check. Did they eat the food you left last time? Are they moving correctly? Shine a flashlight on them to see if they move into darkness quickly. This all indicates healthy worms.

Do a sniff test. If things stink, you have a problem. Check the troubleshooting guide below for possible causes.

After a couple of months, you can check for cocoons. They are the size of small rice grains and not easy to spot. Seeing them is a good sign that the worms are healthy, and you should soon see baby worms.

Check moisture levels with each visit and add more water if needed. Also drain off any leachate before it starts to stink. The ideal moisture content is 75%.

Check the bedding, and if it is packed down, fluff it up again. If the contents of the bin is less than 50% bedding, add some more. You normally have to do this once a month, but this depends very much on the number of worms you have and the amount of food you give them.

One pound of worms (1,000) needs about one square foot of space. As the worms multiply, the farm will reach a point where there are too many worms in the bin. When this happens, their health will decline and you need to do something about it. You can start a second bin, give some away, or sell them. If you put them in the garden, they will likely die, especially in cold climates.

The bin is full of organic matter, including fungi, which may start forming fruiting bodies, i.e., mushrooms. That is not a problem, but it may be best to remove them as they form. Moldy food is normally not a problem, but if the worms are not eating it, remove it from the bin.

Winter Care

Worms that are kept indoors should be fine even if they are in a colder basement, but keep an eye on the minimum temperature. As temperatures drop, the worms will eat less and be less active.

Outdoor worm bins in cold climates need to be brought indoors in winter because the worms won't live through a freeze. In marginal areas, you can leave it outside if you insulate the bin with Styrofoam and cover the whole thing with fall leaves.

Troubleshooting

Worm bins should have a healthy earthy smell. If the odor changes, it is time to have a close look at all of the conditions worms like, to determine the cause.

Symptom	Problem	Solution
Ammonia smell	Nitrogen levels are too high.	Add less high nitrogen food or try to use bedding with a higher C/N ratio.
Rotten smell	The pile has gone anaerobic and needs more oxygen.	Check the moisture of the bedding and dry it out if it is too wet.
	Rotting food.	Remove old food.
	Not enough oxygen.	Increase venting and fluff up the bedding.
	Old leachate.	Remove leachate.
Too warm	Bedding is composting too fast.	Use bedding with a higher C/N ratio.
	Too many worm castings are decomposing.	Remove compost and castings.
Worms leaving the bin	Worms are stressed.	Turn on a light and get them to stay in the bedding until they settle down.
	Problem with the bin or contents.	Worms may leave if any other issue is causing a problem. Fix the problem and worms should settle down.
	Bin not dark enough.	Move the bin to a darker area, or decrease the light.
Worms not eating	Could be any of the other issues.	Go through a health check to determine the problem.
Invading insects	Insects around the food.	Use a screen with smaller holes and reduce the amount of food.
No cocoons or baby worms	Worms are not mature enough or not settled in a new bin.	Give it more time.
	Low sex drive.	Try withholding food for a week.

Harvest Castings

In a commercial setting, which is maximized for productivity, usable vermicompost is ready in as little as six weeks. In worm bins receiving minimal management (most home systems), vermicompost is ready in four to six months. It is not quite clear how the term "usable" is defined.

You have been taking care of your worm farm for a while, and castings are starting to build up. How do you separate the worms from the vermicompost and the worm castings? This can be easy or difficult depending on your setup.

Before we discuss ways of harvesting castings, it is important to know when you should harvest them. Remember that fresh castings are like fresh manure, they are best if aged for a while. If you harvest too early, you will have few castings and a lot of bedding material. If you wait too long, you will have more castings, almost no fresh bedding, and a lot of composted material. This latter case sounds ideal, but it is not a healthy place for worms to live. You want to harvest at a midway point, where you get a lot of castings but the bin is still a healthy place to live.

The best harvest time is complicated by the amount of bedding material you add after starting the bin. Some people add it all at the beginning of the process, and when most of it is gone, they harvest. Others add some with each feeding. If you do that, there will always be some bedding.

Aim for a point where about 50% of the bin is castings and the remaining is bedding and compost. You might have a hard time telling the difference between castings and compost.

If you prefer, you can go by time and harvest castings about four months after starting the bin.

What does harvesting mean? For the most part, harvesting is the separation of worms from everything else. If the bin contains a lot of uncomposted bedding, you also want to remove some of that. What remains is the vermicompost.

The easiest way to harvest the compost is to select a bin that automatically does this for you, like the stacking trays, or the long bin. But if you are not using one of those, the following options can be used.

Stop adding bedding a few weeks before harvesting and stop feeding a week before. That way all of the food should be eaten. Then use one of the following methods to separate the worms.

Direct to the Garden

This is the easiest method. Take out half of the material in the bin, including worms and compost, and put it in the garden. Add more bedding to the bin and keep the farm going.

Depending on your climate and time of year, the worms may or may not survive, but if they die, they add extra organic matter to the soil.

Hand Sorting

This method is tedious, but it does work. Dump the contents of the bin on a plastic sheet and pull out the worms one by one. Place them in a clean bin to start a new farm.

Bag Method

Find a mesh bag that has holes which are large enough for worms to crawl through. Place some favorite worm food in the bag, like apples, melons, or kiwi. Place the bag into the worm bin and wait for a couple of days. The worms will fill the bag while getting at the food. Simply remove the bag, and you have separated the worms from the compost.

You might need to do this a couple of times to get all or most of the worms.

This is also a good way to collect worms for a friend without disturbing the bin.

Light Method

Shine a light on the bin, wait a few minutes. The light drives the worms deep in the bedding, and then you can harvest the top inch or two of compost. Wait a few more minutes to drive the worms deeper and remove more compost.

Once you have removed most of the material, add fresh bedding and keep the bin going.

Side-to-side Method

Move everything to one side of the bin. Insert some type of divider in the middle of the bin so the compost and worms are on one side and the other side is empty. Place fresh bedding on the open side and some food at the far end away from the old compost. Wait for the worms to migrate from the old side to the new side. Then remove the compost on the old side.

It is important that the divider you use does not completely prevent the worms from getting to the other side.

Sifting Method

Make yourself a sifting screen using one-quarter-inch hardware cloth. Place some of the material from the bin on the screen and shake it. The compost should pass through the screen and leave the worms behind. This sifting process works best if the compost is reasonably dry.

Sifting is very stressful on the worms.

Using Vermicompost

When is vermicompost finished? As discussed in previous chapters, regular hot composting is never really finished, but gardeners call it finished when they can no longer see kitchen scraps in it. We have the same situation with vermicomposting. Many users think it is finished when they harvest the castings, but there is no real "perfect" time to do this. Some will harvest earlier and others later. Each method works fine provided that you use the compost correctly.

A complication is that worms produce castings over a several month period. The oldest ones are well composted by the time the last ones are produced. So the compost is a mixture of material at different stages of decomposition, and just like any other compost, it will continue to decompose for years after being added to soil.

Most gardeners who do vermicomposting talk about "harvesting the castings" and "using the castings," but what they are really talking about is the compost.

Harvested compost is organic material that is partially along the road to being finished compost, but it is years away from that end point.

It can be used right away in the garden, which will reduce the loss of nutrients. However, it should be aged a couple of months before using it for potted plants or in containers. At that point it can be mixed in to potting soil (20% by volume) or layered on top of the soil.

Chemical Composition of Vermicompost

The chemical composition of vermicompost depends on how it is made, the bedding used, and the food that was used. The nutrient levels are very similar to traditional compost. They have the same pH, and they have similar NPK values. The following data comes from a study[1] that used the same input material for both systems.

- NPK for hot compost is 1.4-1-1.3
- NPK for vermicompost is 1.8-3.8-1.3

The slightly higher nitrogen levels are beneficial to gardens, but the higher phosphorus is not. High phosphorus levels are toxic to plants and mycorrhizal fungi. Most gardens have enough phosphorus, and some that have been heavily fertilized with either organic or synthetic products, including compost, show high phosphorus levels.

Storing

Storing vermicompost presents a bit of a problem. The fresh material is moist and contains a lot of microorganisms. In order to keep them alive, it needs to be kept moist, but if you do that, it gets moldy, which most people don't like. If you dry it out, many of the microbes will be killed.

Some believe the microbes are an important contribution to the health of soil and plants, but that is a myth. As discussed in the section called The Microorganism Myth in Chapter 2, adding microbes

Parameter	Vermicompost	Compost
Carbon	24%	23%
Nitrogen	2.1%	1.8%
Phosphorus	1.2%	0.9%
Potassium	1.5%	1.2%
Calcium	2.0%	1.8%
Magnesium	0.8%	0.7%
Copper	57 ppm	31 ppm
Iron	412 ppm	306 ppm
Zinc	89 ppm	61 ppm

Note: Nutrients in vermicompost and garden compost made from vegetable and kitchen waste (dry weight)[2]

to soil does not increase the total number in soil. The real value of vermicompost is the nutrients and organic matter.

If you accept the above statement, then the best way to store the material is to dry it and then store it. Once dry, it will store for a long time.

If you are not convinced and you want to preserve the biology, then use the product quickly. It can be stored for a couple of weeks in any container that allows some air flow and prevents total drying.

Using Leachate

Worm leachate is the liquid that drains out the bottom of the worm bin. This material has created lots of heated debates in the vermicompost community. Some feel very strongly that if you have leachate, you are growing your worms too wet—it is a sign of poor culture. Others get leachate all the time and use it to feed plants. There are even commercial producers who make leachate for sale.

Producing leachate is not a sign of a poorly run farm. Provided it is able to drain off regularly, it does not harm the worms. On the other hand, it's fine if you don't get any. It just means your bin is a bit drier and you are feeding food that has less moisture in it.

What is leachate? It is a dark liquid that has filtered through the contents of the bin picking up all kinds of chemicals. Many of these are nutrients, and that is why it is good for plant growth. It also picks up microbes which will also feed plants as they die and decompose. Think of leachate as concentrated plant juices full of sugars, proteins, other carbohydrates, and minerals.

Because it is a liquid, it tends to be anaerobic and starts to smell quickly. For this reason, it should be drained off every few days and used within a few days. Here are several ways to use it.

Compost Myth: Worm Leachate Is Toxic and Contains Pathogens

Many gardeners believe that worm leachate contains phytotoxins that will harm plants. The other concern is that it contains pathogens because it is anaerobic. Neither of these is true.

Does leachate contain phytotoxins? Leachate is created as water runs through the worm bin and collects at the bottom. If leachate contains phytotoxins, it means the vermicompost in the bin also has phytotoxins, and yet vermicompost is considered to be one of the best plant foods.

Leachate probably contains some phytotoxins, as does vermicompost and all other types of compost. Even the soil contains some. But the levels are all so low, they don't kill plants. If you have a concern, try the Seedling Test with your diluted leachate.

Many gardeners share a common belief that anaerobic material is full of pathogens which makes it harmful to us and plants. I think this belief is based on the fact that anaerobic material tends to stink. Anaerobic environments do contain pathogens, but these are also found in aerobic conditions. Any pathogens in the leachate came from the compost in the bin.

When scientists checked for human pathogens in worm leachate, they did not find levels higher than other places, like soil. Besides, human pathogens won't harm plants.

- Pour it back in the bin if the bin is getting dry. This keeps the bin moist, and the nutrients will help feed microbes in the bin. Some even report worms like eating it.
- Dilute it with water, 10:1. Since everybody's leachate is different, these are approximate dilution ratios. Diluting it more won't hurt. Then use it to water houseplants or outdoor containers. If it is too strong, it could burn plant roots in the same way as strong fertilizer.
- Pour it directly in the garden. This does not need to be diluted provided you don't pour it right on plants.

Vermicompost Tea

Vermicompost can also be used to make tea, just like any other compost. Compost Tea is discussed in Chapter 12.

Is Vermicompost Special?

Vermicompost is touted as very special compost that is much superior to other forms of compost. There is a mystique about it and a cult-like following that believes and promotes its superior qualities. Much of this is misplaced.

The nutrient levels of vermicompost are in the same range as other composts, and nutrients are the main value of any compost.

There is a lot of talk about the microbes in compost, but to be honest, they have not been studied very well. Most microbes in soil or compost have not even been identified yet. It is estimated that soil contains a billion species of bacteria, and we have identified 30,000. I have discussed this in more detail in my book *Soil Science for Gardeners*.

When microbes are added to soil, they tend to die and have little if any effect on the native population.

The pathogen level in hot compost and vermicompost is about the same since worms reduce the pathogen level through their digestion system.

The other important component of compost is carbon. It is critical for supporting soil life and improving soil structure. When hot compost and vermicompost are made from the same manure and then compared, the final C/N ratio was lower for vermicomposting, indicating that it was a more complete process. CO_2 production is higher during vermicomposting than conventional aerobic composting, also indicating that the end product is more decomposed, but it also means vermicomposting adds less carbon to soil.

What about growing plants, that's what really counts? When vermicompost is compared to hot compost, they both produce about the same plant growth.

Imagine that you compost an old rotten pepper. It will release about the same nutrients and carbon levels no matter how it is composted. One method might be faster than another, but you only have the elements that are in the pepper. Composting can't make more nutrients.

Some will say that passing the pepper through a worm is somehow magical and that the worm adds value to the process, but keep in mind that, besides grinding, much of the digestion taking place in a worm is the result of bacterial digestion, just like a compost pile.

All of the composting methods in this book depend on microbe digestion. The various methods simply manipulate the microbes a bit differently, but the end product is the same or very similar, no matter which method is used. The material produced by vermicomposting is not any more special than any other method.

Vermicomposting

Pros	Cons
Can be done in small or large spaces.	Requires more time.
No turning.	Need to buy worms.
No concerns about C/N ratio.	Seeds remain viable.
May retain more nitrogen than hot composting.	Worms need to be separated from the compost.
Can be done indoors.	Not ideal for handling a lot of yard waste in a home garden setting.

9

Bokashi Composting

Bokashi is a Japanese word that means "fermented organic matter." It is a completely different way of dealing with food scraps and yard waste. Instead of using an aerobic environment, bokashi is done anaerobically, making it quite different from most of the other composting methods described in this book.

The method is called bokashi composting, but it is not a composting method. Bokashi is a fermentation process similar to the process used for making sauerkraut. It is a kind of pickling process but without the use of salt. It would be more accurate to call it a "pre-composting" method.

Bokashi can be done in small batches in the home or as very large batches in a farm setting. The latter is not commonly used by gardeners, so the focus here will be on smaller home systems.

The waste organic matter is combined with bokashi bran, which contains fermentation microbes, and placed in a special bokashi bucket. More organic material is added as it becomes available. Fermentation takes place in the anaerobic environment of the closed bucket.

After a few days, liquid starts to form in the bottom of the pail. This bokashi leachate needs to be drained or it will start to stink. It can be used to fertilize plants either indoors or out in the garden.

Once the bucket is full, it is set aside for a couple of weeks to finish the fermentation process. After that, the material is taken outside and either buried in the garden, added to a compost pile, or used as food in a worm farm. Bokashi produces fermented organic matter called bokashi ferment. This material then needs to be composted before plants can use it.

You might be wondering why you should bother with bokashi. Why not just compost the starting material? Here are some reasons for doing bokashi.

- It is a method that can be done in an apartment that does not have access to a garden. The ferment still needs to be taken somewhere, like a friend's garden or a community composting facility. The latter is becoming more popular in some communities in the Philippines.
- You can also use ferment indoors in something called a soil factory, as discussed later in this chapter.
- Bokashi can be done in cold climates during winter while the compost pile outside has stopped working. The ferment can then be taken outside in spring to jump-start the compost pile. This saves you trudging out in the snow to dump a few kitchen scraps.
- A contact of mine who lives in the far northern Arctic uses ferment as soil because there is no local soil and flying it in is very expensive.
- Bokashi can be used to compost all food including meat, dairy, and eggs.
- The same method can be used for small amounts of waste and for very large amounts.
- It does a better job of preserving nitrogen, compared to other methods.

A common claim for the advantage of using bokashi is that the "effective microbes" are so beneficial to soil—which is mostly a myth that I will discuss below.

The Fermentation Process

Fermentation is a process that is carried out by special microbes in an anaerobic environment. These organisms use sugar and other simple carbohydrates as food and convert them to a range of organic acids, including lactic acid, butyric acid, and acetic acid (vinegar).

This lowers the pH and kills off any microbes that can't live in these acidic environments.

The acidic environment then ferments the organic matter. In effect it is turning kitchen scraps into pickles and sauerkraut.

The whole process is started by adding some bokashi bran. This bran has been saturated with either sugar or molasses (essentially sugar), and some special microbes. The sugar is easy for bacteria to digest, and they turn it into acids. Once the process starts, it is self-perpetuating. The acidic environment only allows the right organisms to grow, and the liquid produced helps keep the environment anaerobic.

New organic matter added to the pail is quickly fermented.

At the end of the process, the material looks pretty much the same as when it was added. An orange looks like an orange, and an apple is still an apple. It might be a bit softer, but at the end of bokashi, there has been limited chemical decomposition. Fermentation is a preservation process, and the resulting material is quite stable as long as oxygen is kept out.

Fermentation vs Composting

The microbes in compost use the energy in the organic matter and produce CO_2. The amount produced depends on many things, including the C/N ratio of the starting material, temperature, and duration. The released CO_2 is not good for the environment.

During fermentation, very little of the organic matter is decomposed, so it produces very little CO_2. The liquid that drains to the bottom of the pail is mostly water from the food scraps. At first glance, the lack of produced CO_2 makes this system seem to be a more eco-friendly option, but it is important to remember that the true composting process happens after bokashi. It is not clear if the whole process is more eco-friendly.

Composting does lose nitrogen as ammonia and N_2 gas, and as much as 25% to 75% can be lost. Very little nitrogen is lost during bokashi.

Composting is most efficient when the C/N ratio is around 30:1 while fermentation works best at 10:1. Kitchen scraps are perfect for fermentation, but most yard waste has too much carbon. Manure is traditionally added to the yard waste to increase the nitrogen level and lower the C/N ratio before the bokashi process is started.

From a nutrient point of view, ferment is better than compost because it contains more of the original nitrogen. However, it is important to understand that the concentration of nutrients is higher in compost. This might seem like a contradiction but it's not. Ferment weighs about the same before and after the process. Compost, on the other hand, weighs much less mostly due to a loss of carbon and oxygen, but the nutrients, except for nitrogen, have remained the same. This means the nutrients are more concentrated in compost.

The pH of compost is around 7 or slightly above, and ferment is around 4. This acidic nature of ferment will not likely affect soil pH unless it is sandy soil or very large amounts are added. (Keep in mind that soil has a great pH buffering capacity.) The acidic condition in fermentation kills off most pathogens.

A common claim is that the ferment from bokashi decomposes quickly once it is added to soil. This is based on the fact that it "disappears" quickly. Decomposition is a chemical process whereby large molecules are converted into small molecules. There is no way to judge the extent of decomposition by "looking at it" except in the very early stages. Fermentation produces mushy material that may simply fall apart when mixed with soil, to a point where we no longer see bits and pieces. That does not mean the material has decomposed.

I have yet to find a study that looks at the rate of decomposition once ferment is added to soil, and so I am not convinced it decomposes more quickly.

The Bokashi Method

The process is started by placing a small handful of bokashi bran in the bottom of the vessel. You only need enough to cover the bottom. Next, place a layer of food scraps on top of the bran, and cover that with some more bran. Then squish this down to expel as much of the air as you can. Close the lid and set it aside. There will be very few if any odors with the lid closed.

When you are ready to add more kitchen scraps, open the vessel, add the food and some more bran. Squish it down and put the lid back on. People use a variety of squishing tools. Your hands work fine if you don't mind getting them mushy. The masher used for making mashed potatoes works great.

How much bran do you add? The bran is important to get things started, so a bit extra at the beginning is a good idea. You do not need to add as much bran once the vessel is partially filled because the microbes are already working in the rest of the bucket.

Add extra bran if you add meat or dairy since they are harder to ferment. If you add a lot of organic matter at one time, add bran between each two-inch layer of food. Chopped food ferments quicker than large pieces.

Many bokashi buckets have a spigot at the bottom to drain off liquid that is formed during the process. Check for bokashi leachate every few days and remove any that has formed.

Some people add a piece of plastic on top of the material in the vessel to prevent oxygen getting to the food, but this should not be necessary if the vessel is airtight. Store the vessel out of direct sunlight and somewhere warm, between 60°F and 90°F (15°C and 38°C). Don't open the vessel too often to add scraps. It is better to store the scraps outside of the bokashi vessel and only add them every few days.

Once the vessel is full, close it up, mark the date on it, and set it aside for two weeks to complete the final fermentation process. Keep removing any leachate that is formed. It is then ready to be used.

Bokashi Bucket

Bokashi is carried out in specialized buckets, although they can be quite simple, provided certain requirements are met. They need to have a tight-sealing lid, as well as a way to drain off the bokashi leachate.

Commercial products have a false bottom that has holes in it so that the leachate can easily drain to the bottom of the vessel where it is removed using a spigot. Most of these buckets hold five gallons of ferment.

Bokashi kit containing a bucket and bokashi bran mix.

A simple DIY solution uses two nested buckets. The bottom one is used as is and does not require a lid. The upper one has one-quarter-inch (6 mm) holes drilled in the bottom and has a tight-fitting lid. Food and bran are placed in the upper bucket. When it is time to drain off the leachate, the upper bucket is lifted out of the lower one and the liquid in the lower one is poured off. Then the upper bucket is again inserted in the lower one. You could add a spigot to make draining easier, but these tend to leak on round buckets.

One of the benefits of bokashi is that it can be scaled easily. A two-gallon bucket is treated the same as a 100-gallon bucket, although it is a good idea to fill the bucket in a reasonably short period of time of two to six weeks. Large buckets with a large amount of air space hold too much oxygen and have trouble becoming anaerobic.

Bokashi Bran

The bokashi bran contains three important ingredients: bran, microbes, and sugar.

The bran is mostly a carrier to hold the other two ingredients. Wheat bran is the most popular carrier, but almost any fine grain material will work, and rice bran is commonly used. Other options include wheat mill run (a waste product from flour milling), dried leaves, peat moss, sawdust, and some even use shredded newspaper, although the latter may not work as well.

The microbes are responsible for the fermentation process. Adding them to the food ensures that the right microbes are there. When they are exposed to moisture from the food scraps, they start consuming the sugar, which causes a population explosion. The bucket is now full of fermenting microbes ready to tackle the food.

The most common sugar source is molasses, but any sugar will work. From a bacteria's point of view, one source of sucrose is the same as another.

Premade bokashi bran is commercially available, and it is a good idea to use such a product for your first few batches. Once you are more familiar with the process, you can consider making your own to save a bit of money. A method for this is described below.

Troubleshooting

The material in the bucket at the end of the bokashi process should smell like sweet vinegar, a mixture of pickles and yeast. If it has a putrid odor, the fermentation process did not work. You can still add it to a compost pile or bury it in soil. It will slowly compost.

Symptom	Problem	Solution
Green or black mold	The wrong kind of microbes are growing.	Add more bran on top of the mold. Make sure the lid is tightly closed.
	Too much oxygen.	Use a better-fitting lid. Place a plastic bag right on top of the material in the bucket. Open the bucket less often.
White mold	This is not a problem; it is common to see white mold.	

Symptom	Problem	Solution
No white mold	This is also not a problem. It is not always easy to spot.	
Contents of the bucket get too warm	Too much nitrogen.	Add some more high carbon material, like shredded newspaper or brown leaves.
No fermentation—foul odor	Not enough LAB (see next section).	Use more bran.
Bokashi leachate smells bad	The wrong kind of anaerobic microbes are growing in it.	Remove it more frequently.

The Power of Effective Microbes (EM)

Dr. Higa, the person who originally developed the bokashi system, also developed a special blend of microbes called "Effective Microorganisms" (EM). All kinds of special properties have been attributed to this mixture, and bokashi is just one use for them.

Effective Microbes are a mixture of approximately 80 different microorganisms that are capable of positively influencing the decomposition of organic matter. EMRO (EM Research Organization) was the first company to make and promote the use of a variety of EM products. Their main product is EM-1 and made for general use. A special version used for bokashi is called EM-1 Bokashi and consists of fermented rice bran or wheat bran inoculated with EM-1. Another product, called EM-1 Compost, is fermented organic matter inoculated with EM-1 and is used as a starter for traditional composting.

The specific formulation of EM-1 is a trade secret, but some of the key organisms in it are known and they include the following.

- Lactic acid bacteria (LAB): *Lactobacillus casei*
- Photosynthetic bacteria (purple non-sulfur bacteria or PNSB): *Rhodopseudomonas palustris*
- Yeast: *Saccharomyces cerevisiae*

A number of other companies now supply an EM mixture, and there is a lot of information online for making your own concoction.

Lactic acid bacteria are the main organism responsible for starting the fermentation process. These friendly bacteria are an important part of your digestive system and are also found in your urinary and genital tracts. They are also one of the probiotics and the key bacteria for making sauerkraut. These bacteria are found on skin, in all types of soil, and on plant material.

People performing home bokashi have been conditioned to believe that bokashi won't work without EM. In agriculture, bokashi is sometimes made with EM, but a lot of the time, it is made with indigenous microorganisms (IMO—see below), as well as other local microbe concoctions. One study found no difference between using IMO, yeast, or nothing at all. Some people also make their own bokashi bran using IMO that they collect and grow themselves. Others use competing products to EM-1 that contain different organisms. Dr. Higa is said to have always stressed that "it isn't the exact combination or ratio of microorganisms that makes EM so powerful, it is the fact that the microbes are working together as a group."

We know EM works, but it is quite possible that you don't need it, or at the very least, you can substitute other organisms. If you are a science type, this would be a good place to run some experiments.

The bokashi fermentation process sounds almost identical to another process called ensiling, or silage fermentation. In it, feedstock is chopped up and put into airtight silos where the material ferments and preserves the food, which is later fed to farm animals. This process has been used for hundreds of years and does not require special EM organisms. Instead it relies on the organisms that are naturally present on the organic matter.

The key organism in silage fermentation is *Lactobacillus casei*, the same bacteria used in bokashi. It makes use of natural sugar in the feedstock, and enough of it lives on the plant material to start the whole fermentation process.

What is clear is that the added sugar is an important ingredient for getting any fermentation process started.

Brew Your Own EM Microbes

I suggest that you use a commercial bran product that contains EM for your first couple of batches of bokashi. After that you can consider making your own microbe cultures.

I mentioned indigenous microorganisms (IMO) above. These are a group of local native microbes that inhabit the soil and the surfaces of all living things. You can collect your own quite easily.

Cook some white rice in water. Form the rice into two-inch (50 mm) diameter balls. Place these in the soil and barely cover them. If it is cold outside consider doing this indoors using old potting soil. After about three days in warm weather, the rice will be covered with white, yellow, and green IMO. Scrape this off and place it in a jar. Don't worry if you add a bit of rice at the same time. Add dark brown sugar at a ratio of one part sugar to three parts IMO. Let it sit for three days. Now add it to a 50:50 mixture of soil (can be potting soil) and corn flour (or corn meal, bran, or other starchy material). Add water to moisten the mixture and keep it moist until you have used it all. Use this in place of bokashi bran.

There are several other recipes online. Search for "DIY indigenous microorganisms."

Making Bokashi Bran

You can buy bokashi bran or make your own. The following recipe is taken from the book *Bokashi Composting* by Adam Footer, which goes into much more detail about bokashi than I do here.

You will need the following ingredients.
- 12 lb. wheat bran or rice bran
- 3 oz. EM mother culture (or your own IMO)
- 3 oz. blackstrap molasses
- 128 oz. water

Boil the water and molasses and let it cool. Add the EM mother culture followed by the bran. Mix it all together. It should be moist enough so that you can form it into a ball without it breaking apart, but not so wet that water drips out if you squeeze it. Add more water or bran to adjust the moisture.

Take a garbage bag and put all of the mixture in it. Squeeze out as much air as you can and then seal the bag. Put the first bag inside another bag and seal it as well. Set the bags aside at room temperature for two weeks to allow fermentation to take place. Don't open the bags or you will let air in.

After two weeks, open the bag and check on the bran. It should have a sweet, sour, yeasty smell very similar to a winery. A musty, moldy smell indicates that fermentation did not take place. Seeing white mold is normal, but green or black mold indicates a problem. If the bran did not ferment, don't use it for bokashi. Instead add it to a compost pile.

Take the bran and spread it out so that it can dry. Periodically, break it up and mix it up so that all parts are exposed to air. Once it is fully dry, it can be stored and will keep for years, provided it remains dry.

Using Bokashi Leachate (Tea)

Excess liquid is drained off during the bokashi process, and this is called bokashi tea or bokashi leachate. I'll use the latter term to be consistent with my use of leachate for vermicomposting and so that this material is not confused with compost tea.

The leachate should be removed every couple of days because it will quickly start to smell as microbes grow in this anaerobic environment.

Bokashi leachate is promoted as having very high nutrient values and thus is a good fertilizer for plants. The following table provides nutrient values for pure bokashi leachate (concentrate), diluted bokashi leachate (1:50 dilution), and values from a good fertilizer solution developed by Michigan State University (MSU). The bokashi was made from kitchen scraps.

Bokashi leachate is commonly diluted, which will provide the levels in the middle column of the table. The nitrogen level is very

	Bokashi leachate (concentrated)	Bokashi leachate (diluted 1:50)	MSU fertilizer (recommended strength)
Nitrogen (total) mg/l	295	6	68
Phosphorus, mg/l	960	19	7
Potassium, mg/l	4300	86	62
Calcium, mg/l	2500	50	42
Magnesium, mg/l	370	7	10
Sodium, mg/l	24	0.5	0
Chlorine, mg/l	85	1.7	0
Iron, ug/l	2700	54	880
Manganese, ug/l	920	18	420
Zinc, ug/l	3000	60	200
Copper, ug/l	<200	<4	200
Boron, ug/l	1700	34	50
Molybdenum, ug/l	<200	<4	50

Credit: Based on research by Håkan Asp

low, and the phosphorus level is too high. Most micronutrients are on the low side. It also contains high levels of sodium (24 mg/l) and chloride (85 mg/l), both of which are toxic to plants at low levels. Sodium is toxic at 50 mg/l, and chloride is toxic at 70 mg/l.

The amount of sodium and chloride depends very much on the type of material used. Much of the sodium chloride would be from salt added during cooking or salt added as a preservative in commercial food.

A research study[1] looked at growing pak choi, using household bokashi leachate, and compared it to regular fertilizer. Growth with bokashi leachate was much slower, probably due to a lack of nitrogen.

A   **B**   **Control**

Test plants were treated with bokashi leachate from a school
kitchen (A) and a home (B). The control was treated with
commercial fertilizer. Based on research by Håkan Asp,
Swedish University of Agricultural Sciences.

Bokashi leachate is also acidic, so it should not be poured onto
plants. Remember, some people use vinegar to try and kill weeds.

Bokashi leachate is not a great fertilizer for plants. I would not
use it on house plants or plants in outdoor containers, but it can be
used in the garden after dilution (1:100) to reduce the sodium level.

Using Bokashi Ferment

Bokashi ferment is the material left in the bucket at the end of the
process.

The kitchen scraps still look like kitchen scraps. You can see the
orange peels and the bread. It is a bit moldy and it is softer, but for
the most part, there has been very little decomposition so far.

What do you do with this ferment? It can be composted in a
variety of ways, or used to improve soil in a "soil factory," as described in the next section, or it can be used as feedstock in your
worm farm.

If you are doing bokashi in an apartment where you don't have a
garden or worm farm, you have a bit of a dilemma. You either have
to put it in the garbage, which defeats the whole purpose, or give it
a friend for their garden. A few countries are now setting up collection sites for the material.

Composting Bokashi Ferment

Bokashi ferment can be added to any of the composting methods described in this book. If it is coming from your kitchen, it will have a C/N ratio of about 30.

Another popular option is to bury the ferment using the trench composting method. It is claimed that bokashi ferment composts more quickly in a trench than the same unfermented material, but I have yet to see any research to confirm this. To compare apples to apples you have to start the clock at the beginning of the bokashi process. I suspect that when you add the time to do bokashi to the time of trench composting, both methods will come out about the same.

If you put it directly in the soil, don't put it too near plants and don't plant right on top of it for at least two weeks. This allows time for soil bacteria to digest the acids.

Food for Worms

There is a lot of controversy about feeding bokashi ferment to worms. Some say it will harm them, others say the worms really like it.

The concern for worms may stem from the myth that acidic food will harm them. It is true that they like living in a more neutral pH, but feeding them acidic food like citrus fruit will neither harm them nor make their environment more acidic, provided that the amount added is reasonable.

The same is true for ferment. Some people will add small amounts at a time, especially when first introducing the food to worms. Once they get accustomed to it, more can be added with each feeding. As long as they eat it all, it won't harm them nor make the bin acidic.

I am left with a question. If you plan to feed your worms after making bokashi, why not just feed the food directly to worms and skip the bokashi process? The only reason I can think of is that bokashi is better at handling meat and dairy—but how much of that do you really generate? Meat and dairy are too expensive to throw out.

Soil Factory Using Bokashi Ferment

I am seeing the term soil factory used more often, usually in discussions about bokashi or vermicompost. I'll focus here on bokashi, but most of the comments also apply to vermicomposting.

Soil factory is a process where bokashi ferment is turned into "super" soil, or so it is claimed. Let's clear up one point right away. This process does NOT MAKE SOIL! It is really a form of indoor trench composting, and nothing more.

The process is started by placing some soil in a container. It could be garden soil, potting soil, or even old soil from potted plants. Bokashi ferment is added and topped with more soil. You can mix this up or just leave it. In a few weeks or a month, the ferment has disappeared and you are left with improved soil.

This method is promoted by gardeners, but I have yet to see anyone test this new soil to determine how super it really is. There is no doubt that some compost and nutrients have been added to the soil.

The problem I see with this process is that people think bokashi composts quickly and becomes part of the soil, but it doesn't. Composting is a slow process and takes months and years, not a couple of weeks.

Is this a good use of ferment? There is no doubt that nutrients have been added to the soil. One problem is that no one has tested this soil to see how nutritious it is. It might have a high C/N ratio, in which case microbes will take nitrogen away from plants so they can digest the carbon and then plants suffer. This will depend very much on the material added to make the bokashi. If you add a lot of starchy food and newspaper, it will be a bigger issue.

Are there other benefits, like added microbes or better soil structure? Probably not, especially with potting mixes, which are not soil to begin with.

A New Type of Soil Factory

While experimenting with this process, I decided to try an improved soil factory method.

Take your bokashi ferment and homogenize it in a blender to make a smoothy. Then pour this into the soil. Mix it up and you are done. Instant fortified soil with no waiting period.

I call this new method the Instant Soil Factory. But, so far, I have no idea if it is any good for plants.

Pros and Cons of Bokashi

I am still not sold on bokashi. The leachate has limited value, and the fermented food scraps still need to be composted or disposed of. If you are going to dispose of them in the garden, you might as well compost instead. I do see advantages for apartment dwellers or for people with limited access to soil.

Bokashi may have some environmental advantages over other composting methods, as described later in this book. If this turns out to be true, it may be a better process for gardens, but then it requires the use of larger vessels, or other ways to keep large piles anaerobic.

One of the main claimed benefits of bokashi is that the "finished product is full of beneficial microorganisms." That is not true. Most of the organisms are *Lactobacillus* bacteria, and they are quite ubiquitous. Adding them to soil is of no value because they are already there.

Bokashi

Pros	Cons
Composts meat and dairy products.	Need to add bokashi bran.
No nutrient loss.	More expensive than composting.
No turning.	The ferment still needs to be composted.
No need to adjust the C/N ratio.	Bokashi leachate is of limited value.
Suitable for indoors.	Not a good mulch option.
Can be done on any scale.	Works best with chopped up food.

Another key claimed benefit is that bokashi is quick—it only takes two weeks. That is not true either. Most homeowners will take about two weeks to collect enough material to fill the bucket. It then sits for another two weeks to ferment. After that it is added to the garden where it takes a month or more to reach a point where the material is no longer recognizable. It may be faster than hot composting, but there is almost no data to show this.

10
Eco-enzyme (Garbage Enzymes)

Eco-enzyme is a similar process to bokashi for composting kitchen scraps. It can be done inside using small batches, so it is ideal for homeowners. It is not a popular process in North America, but it is used a lot in Asia, particularly Thailand.

It is also called garbage enzymes since it was started as a way of reducing garbage going to landfill. The resulting liquid can be used as a cleaning product or to fertilize plants. The fermented solids can be added to a compost bin or directly to soil.

What is Eco-enzyme?

Eco-enzyme is a fermentation process of organic waste. Any plant-based kitchen scraps can be used, and some people focus on fruits and citrus rinds.

The process is very similar to making sauerkraut. At the end of process, you end up with fermented organic matter—the kraut and a liquid component, the eco-enzyme tea. The two components are separated and used individually.

Unlike bokashi, there are no added microbes. The process relies on the native microbes already on food scraps. You can't see them, but if you looked at the scraps with a microscope, you would see that each piece is covered with billions of bacteria, fungi, and yeast. These organisms will carry out the fermentation.

Fermentation produces several organic acids, including lactic and acetic, which lower the pH. The anaerobic conditions prevent pathogens from growing.

This process is called eco-enzyme because the microbes produce a number of enzymes for digesting the food, and it is these enzymes that are responsible for its benefits.

What is an enzyme?

Enzymes are special proteins made in all living organisms, and they carry out most of the chemical reactions that take place in cells. They build the molecules needed to make cells, and they break molecules apart during the decomposition process. All of the internal processes, like photosynthesis, are also carried out by enzymes.

Bacteria don't have mouths although some can encapsulate small pieces of organic matter. Most get their food by excreting enzymes. The enzymes digest the organic matter around them, including other living microbes. Once the digested molecules are small enough, the bacteria can absorb them through their cell membranes.

Enzymes are just molecules—they are not living. When added to soil, they will help decompose organic matter, and through that process, they can provide nutrients for plants. Some enzymes digest other enzymes so they are relatively short-lived in soil.

The claimed benefits of eco-enzyme are due mostly to the enzymes produced during the fermentation process.

Making Eco-enzyme

The organic matter is chopped up into small pieces so they ferment quicker. Any fruits and vegetables work well, and some prefer to use mostly fruit peels. Things like meat and bones should not be used. Eggshells are not used because they will neutralize the acids, thereby slowing down or even stopping fermentation.

To start the process, take one part brown sugar or molasses, three parts organic matter (kitchen scraps), and ten parts water. Place this in a closed bottle and let sit for three months at room temperature. Open the lid once a week to allow excess pressure to escape. At the end of the process, pour off the liquid (eco-enzyme tea) and separate it from the solid material left in the bottom of the bottle (ferment).

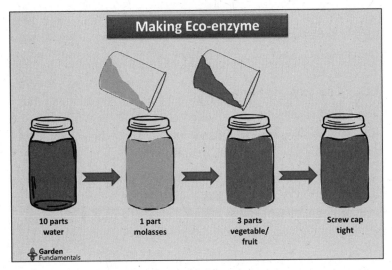

Recipe for making eco-enzyme.

Using Eco-enzyme Tea

The eco-enzyme tea contains lactic and acetic acids, alcohols, various bacteria, and yeast. It also contains enzymes including protease, amylase, and lipase, as well as plant nutrients.

The acids, alcohols, and enzymes give the tea antifungal and antibacterial properties, and it is used for cleaning household surfaces. For tough jobs, it is used neat, and for less demanding tasks, it can be diluted with water to a 1:10 or even 1:50 mixture.

Here are some suggestions for using the tea.

- Clean toilet bowls
- Add to your bath to soften skin
- Add to the washing machine to soften clothes
- Deodorize shoes
- Remove black mold
- Sterilize countertops

The tea can also be diluted to a 1:100 ratio and used as a plant fertilizer. How cool is that—shine your leaves and feed the plant at the same time.

There has been little scientific testing of these teas, but I would expect the amount of nutrients to be similar to bokashi ferment (i.e., tea).

Using Eco-enzyme Ferment

The solid ferment that is produced is similar to bokashi ferment. It is not yet decomposed, so it needs further processing in order for it to be useful to plants.

You can simply bury it in the garden where it will complete its decomposition or add it to another composting system.

11

Buying Compost

Compost is great for gardens, and even if you don't make enough on your own, there are other sources.

Municipal Compost

Many municipalities are now offering a green bin system that allows them to pick up organic matter and compost it in a central location. The finished compost may be available to residents, either free or for a small fee.

Each municipality will have a different way of processing the green material, but in general they use a hot composting process that is highly controlled.

National quality standards have been developed for many European countries, but in North America this is controlled by provincial and state regulations. There are no national standards.

Municipal compost is safe for general garden use, but there are some specific concerns about its use in vegetable gardens. If you have enough of your own homemade compost, use that for the vegetable garden and municipal compost in the rest of the garden.

Pathogens and Weed Seeds

The hot composting process will kill pathogens and weed seeds. Many municipalities also now allow pet waste in the green bin because their systems will any kill pathogens in it.

Plastic

Some facilities collect yard waste in plastic bags. The facilities then rip the bags apart and try to separate plastic from organic matter. This process is not 100% effective, and small plastic fragments end up in the compost.

Other municipalities now use a green bin and ban plastics from being added to it. The resulting compost is much cleaner and contains almost no visible plastic pieces.

Municipal compost contains between 0% and 1.5% (dry weight) plastic consisting of both hard plastic and film plastic (plastic bags). The primary source of plastic contamination in food waste streams appears to be food packaging and containers, most likely from residential, commercial, and institutional sources.

Microplastics

Microplastics are very small particles of plastic that are less than five mm (0.20 inch) in length. They are created as plastic degrades, and composting creates them when larger pieces of plastic are added along with organic matter. In the environment, microplastics keep breaking up into smaller and smaller pieces eventually forming nanoplastic particles, but the process is slow and even very tiny pieces can impact the environment.

A study looking at municipal compost from facilities in Spain[1] found 10–30 particles of microplastics per gram of dry compost.

Food itself is also a source of microplastic particles. The level of plastic contamination present in food waste streams is not well characterized in the scientific literature, but a recent study[2] (Environmental Protection Agency, 2021) found 300 pieces of microplastics per gram of food waste collected from grocery stores in the United States. Even bottled water contained 325 particles per liter. The current research looking at the effect of these small particles on food production and soil is limited. There is no evidence that plant roots absorb microplastics from soil, but they might absorb nanoplastics. Both sizes of particles can affect the properties of soil, including texture and structure, which in turn may impact plant performance. One study[3] that looked at the effect of adding municipal compost containing microplastics on soil during a 9-month period found they had "no significant negative effect on wheat seedling emergence; wheat biomass production; earthworm growth, mortality, or avoidance behaviour; and nematode mortality or reproduction; compared to controls." Other studies using high doses of microplastics[4] found that soil properties were altered and seedling growth was affected.

What is not known is the long-term effects of accumulated amounts in soil. So, should you use municipal compost? Until we have a better understanding of the effects of microplastics on soil, it is prudent to avoid them.

Chemical Contamination

A variety of chemicals may cause concern.

Pesticide residues on food are not a real concern. The amount of these chemicals on food is very small, and most will degrade during the composting process.

Heavy metal contamination should also not be a concern provided that the compost is made from kitchen scraps and yard waste collected from homes. Heavy metals do not decompose and will remain in the compost, but any single source of higher levels will be diluted by all the other sources.

It is important to understand that all soil, and therefore all plant material, contains some level of heavy metals. Except in some extreme cases, the level in soil and plants is well below safe limits.

Other toxic chemicals get into the compost as contaminants on the input ingredients. For example, leaves picked up from city streets often contain oil and other automotive wastes.

The one category of chemical that has been an issue in the past is the presence of specific herbicides. They can end up in compost made from grass clippings, hay, straw, and manures from herbivorous animals that have grazed in pastures or eaten hay or grains containing persistent herbicides.

Not all herbicides are a problem. Some like glyphosate are short-lived and don't pose a problem. Others are digested in the stomachs of animals that eat them. The higher temperature during composting can degrade some to a safe level, and still others will bind to soil, rendering them harmless to plants.

There are, however, a few herbicides that survive all of this and remain in soil for years. These so-called plant growth regulator herbicides (PGR) belong to the pyridine group of compounds and include clopyralid (a dandelion herbicide), aminopyralid, aminocyclopyrachlor, and picloram.[5] The following levels are known to cause harm to plants:

- clopyralid: 10 ppb
- aminopyralid: 1 ppb
- aminocyclopyrachlor: unknown
- picloram: 5 ppb

How common are these in compost? Considering the large number of gardeners who use municipal compost and the small number of reported cases, it is a rare event. However, when it does happen, it is a serious problem. As long as persistent herbicides are on the market, there is a chance that manure, compost, straw, or commercial organic fertilizer can be contaminated.

Plant Diseases

The material collected for municipal composting contains all kinds of diseased plants. This is not really a concern because of the hot composting.

Biosolids (Sewage Sludge)

Biosolids is the fancy name used for sewage sludge which is the semi-solid material produced as a by-product during sewage treatment of industrial or municipal wastewater. It is the solid material that is left after the municipality treats the stuff you send down the drain and toilet.

For many, this material has a strong ick factor, and they would never use it in their garden, but don't be too quick to judge it. It is used as a major source of fertilizer in much of the world.

The treatment of sewage is not very different from the composting processes discussed in this book. You start with a lot of organic matter that is decomposed by a wide range of microbes. The end product is very similar to finished compost.

Consider this, gardeners rave about the great quality of organic matter that is processed through worms or farm animals, but when this same material is processed through humans, it's disgusting. Our digestion system is not that different from a cow or a worm.

A key difference between compost and biosolids is that other ingredients also end up going down the drain, including industrial wastes and chemicals used in the home, like soap and cleaning products. Most municipalities now have fairly strong controls over the type and quantity of industrial waste added to the waste stream.

The main concern for biosolids has been heavy metals, pathogens, and salts. In an effort to make this material more acceptable, it is now tested and classified. Class A biosolids are approved by the U.S. EPA for use in agriculture and food production in gardens, and it can be sold to home gardeners as fertilizer and compost. European regulations are even tighter.

Class A biosolids are tested for fecal coliform, salmonella, and heavy metals. It goes through limited testing for other chemical contaminants.

Class A biosolids are not tested for microplastics, and a recent study in Ontario, Canada,[6] found levels between 9 and 14 particles per gram, which is similar to that found in municipal compost. There was no accumulation in soil, presumably because they leached into lower soil levels and aquatic environments.

Numerous studies have shown that food produced using biosolids is safe to eat, but it is not allowed on certified organic farms. Milorganite is a popular commercial product made from biosolids, and it has an NPK of 6-4-0. It is promoted mostly for use on lawns.

Because of the unknown impact of microplastics, biosolids are not a good source of compost.

Mushroom Compost

The recipe for making mushroom compost varies from company to company, but can include composted wheat or rye straw, peat moss, used horse bedding straw, chicken manure, cottonseed or canola

Compost Myth: Mushroom Compost Contains Table Salt

Mushroom compost is used a lot for gardens, but it comes with a warning: don't use too much because it contains a lot of salt. I have been aware of this warning for a long time, and I have even cautioned people against using too much. Salt can harm plants, and it only makes sense not to use a product loaded with salt.

I always wondered why mushrooms would be grown with high levels of sodium. Are fungi that different from plants? Do they need high sodium levels? A little online research started to make things clear. It turns out to be a good example of the confusion around the term "salt," and is something every gardener should understand.

The general public uses the term salt to mean table salt, which is sodium chloride. This is also used to melt ice in cold climates. As a result, many gardeners know that salt kills plants because they have seen the damage left behind in spring.

Chemists and other scientists use the term salt to refer to any compound that is made up of ions. Sodium chloride is one of many different types of salt. In water, it breaks up into sodium ions (Na^+) and chloride ions (Cl^-).

Compounds such as ammonium nitrate and potassium phosphate, found in synthetic fertilizer, are also salts. Too much fertilizer can damage plants, but these fertilizer salts are not nearly as toxic to plants as sodium.

Unfortunately, gardeners are caught halfway between these two definitions for salt; sometimes it is fertilizer salt, and sometimes it is table salt. When gardeners use the term, you have to stop and ask how the word is being used.

The high salt level in mushroom compost turns out to be a mixture of fertilizer salts and not sodium chloride. It actually contains very low sodium levels.

meal, grape crushings from wineries, soybean meal, potash, gypsum, urea, ammonium nitrate, and lime.

The ingredients are combined and mixed well. It is then left to sit and compost. The ingredients contain a fair amount of nitrogen, so the pile heats quickly, producing a nutrient rich product. When finished, it is pasteurized to kill disease causing organisms and pests.

It is then placed on beds and inoculated with mushroom spawn (mycelium). Once the mushrooms are harvested, the compost can't be used again because it is not nutritious enough for a second batch of mushrooms. However, it still contains lots of nutrients for plants and is sold to nurseries or made available to gardeners.

As it comes out of the mushroom production facility, it is called "fresh" mushroom compost. In most cases it then sits in piles for a few months to further age before it is made available.

Mushroom compost has a pH around 6.6 which is perfect for most gardens and an average NPK of 1-0.7-1. The calcium level is 2.3% which is a bit high, but not a concern in most soils.

This is rich compost and is best used as one-to-two-inch mulch. If you use too much, you can burn plants, but that is true of any manure, compost, or fertilizer.

12

Compost Tea

What is compost tea? This seems like a simple question, but it's not. There is no clear definition of compost, and as you have seen in this book, there are many input ingredients and many ways to make it. Each compost is a bit different.

There are also several ways to make the tea, and what I find amusing is that many people brewing tea feel that their method is the "only right one." All other methods are wrong!

I could write a whole book on compost tea, but that is not the main focus of this book, so I need to keep the information brief. My main focus here is to give you enough background to get you started. I also want to point out a number of myths about compost tea. It does have some benefits in the garden, but unfortunately it has gained a cult-like following that is blindly believing anything someone says.

Making compost tea is fairly simple. Take some organic matter and place it in a pail of water. Wait several days and draw off the liquid—compost tea! As with all composting methods, the process uses microbes to further decompose the organic matter and release nutrients. The environment in the pail also encourages microbe populations to increase. There are two main categories of tea: anaerobic and aerobic.

Anaerobic tea is easy to make. The water does not contain a lot of oxygen and very quickly becomes anaerobic. Without doing anything extra, the material in the pail will be anaerobic, even without a lid.

Aerobic tea, which is also called aerated compost tea (ACT or ACCT), is made the same way except that air is bubbled through the tea during processing. This adds enough oxygen to keep it aerobic.

Types of Compost

Almost anything can be used to make compost tea.

Weed tea is made by placing weeds into water and letting the mixture brew. It can be aerated, or not. It is a great way of getting rid of noxious weeds, but let me warn you that it really stinks if it's not aerated.

Compost tea can also be made from any type of compost.

Manure tea is made from fresh or slightly decomposed manure, and some proponents feel that this makes the best compost tea. You can even buy small tea bag-like sacks containing manure for this use. Talk about expensive manure!

Vermicompost tea is also claimed to be the best.

If you spend some time online in places where compost tea proponents hang out, you quickly realise that each one has a favorite brew and strongly feels theirs is the best. I am talking about gardeners, not scientists. If you pay close attention to the claims, you will see many positive statements, but what you will not see is any real data. They don't measure nutrients in the tea. Nor do they measure microbe types or quantities. The trials they report are missing controls, and without controls you can't reach conclusions about efficacy. To be blunt, they have no idea which tea is the best.

We even have a huge problem in research. Scientists do measure nutrients and quantities of microbes, but there are no standards for the compost. The cow manure in the US is not the same as cow manure in India. How can you compare tea in the two places? It is a huge problem for researchers studying compost tea.

Claimed Benefits of Compost Tea

Making tea from compost adds two main things to the tea: nutrients and microbes.

It is easy to understand nutrients. Compost already has plant-available nutrients in it, and they easily leach into the water while making the tea. Further decomposition in the pail releases even more nutrients. The result is that the tea contains plant nutrients which are good for plants.

The brewing process allows microbes to increase in number. The finished tea has a higher number of microbes than the starting compost, and it is believed that these microbes will benefit the soil.

The increase of nutrients and microbes account for the following claimed benefits.

- Increased plant growth
- Healthier soil
- Disease prevention

Let's look at each of these in more detail.

Compost Tea Increases Plant Growth

It is claimed that compost tea increases plant growth. I have been following the research on this for many years and have looked at many research papers that compare compost tea to water and show an increase in plant growth. No big surprise there.

We know nutrients cause plants to grow better, and we know compost contains nutrients. Ergo, compost tea helps plants grow. I am surprised that millions of dollars have been wasted to prove this again and again.

There is a big hole in this research. There are almost no studies that compare compost tea to synthetic fertilizer containing the same amount of nutrients. This is the only way to show that there are some special compounds in compost tea. I suspect such a test would show no difference. Nutrients in compost tea are the same as nutrients in fertilizer—they will both grow plants the same way.

Compost Tea Builds Healthier Soil

What is healthier soil? I dedicated a whole book to that topic, *Soil Science for Gardeners*, but briefly, it is soil with better structure, a higher amount of organic matter, and a higher population of microbes.

The best way to improve soil is to add more organic matter, and compost is ideal for this. As the level increases, so does the microbe population because it feeds on the organic matter. As the microbe population increases, soil structure improves. It is the life, activity, and death of microbes that aggregates soil and creates better structure.

What does compost tea add to soil? It is mostly water and has almost no organic matter in it. The valuable organic matter is the bulky stuff left in the bottom of the pail. The bit of brown color in the water contains almost no organic matter.

It does contain microbes, but as I explain in the last chapter, adding microbes to soil does not increase the amount of microbes in soil. This is a very common myth in gardening.

Since compost tea does not add any substantial organics to the soil, it follows that it also has limited effect on the health of the soil. You may find this very counterintuitive, and it certainly goes against what others say, but the statement is supported by science.

The best way to improve the health of soil is to take the compost, add it directly to soil and skip making the tea.

Compost Tea Prevents Diseases

Compost tea can be applied to soil, or it can be used as a foliar feed. Both types of applications have been studied for disease suppression.

When microbes are sprayed on leaves in large quantities, they can outcompete the existing microbes and some of these are plant pathogens.

Microbes applied to soil can also have an effect on pathogens, which in turn reduces disease pressures.

There have been studies that show both of these application methods can work in very specific situations. However, there are even more studies that show they don't work. By way of example, a study[1] looked at the effect of both aerated and non-aerated compost tea on the suppression of gray mold and concluded: "The variability in gray mold suppression from NCT and ACT applications indicates that disease control would not be commercially acceptable."

A 2-year study by the Rodale Institute and Pennsylvania State University[2] evaluated the use of aerated compost tea for disease suppression in grapes, potatoes, and pumpkins. They found some suppression of powdery mildew on grape, a slight reduction of gray mold, and an increase in the level of downy mildew. Compost tea failed to suppress powdery mildew on Howden pumpkins and failed to reduce late blight on Superior potatoes. Washington State University[3] has an extensive list of such research.

The big problem here is that each research project starts with different compost. It is possible that one type works and another does not. It is also possible that the results are dependent on differences in the environment, something that is yet to be discovered.

The other issue is that many studies are done in the lab. You grow a plant in a pot and infect it. Then you spray it with compost tea. If you see positive results, you claim success. These studies are certainly valuable, but you can't extrapolate the results to the field. They may not work in the real world. Unfortunately, untrained gardeners or news outlets only grab the headlines and incorrectly conclude compost tea works.

The bottom line is that there is very little consistent science to support the idea that compost tea stops diseases, but there is some evidence it might work in very specific cases. It is certainly not a general pathogen suppressant, as routinely claimed.

What Does the Science Say

I'd like to have a look at one study that provides some interesting insight into the value of compost tea. As I mentioned, most studies compare compost tea to water and see improvements. This never made much sense to me. You have a pile of compost and now have two options: use the compost or make tea and use the tea.

Making tea only makes sense if it works better than the original compost. To study the value of compost tea, you have to compare it to compost, not water, and that is exactly what this study did.

The work was done by Bryant C. Scharenbroch and Gary W. Watson.[4] They looked at the growth of two types of trees, red maple (*Acer rubrum*) and river birch (*Betula nigra*), in compacted urban soil. They wanted to mimic a garden in a new home where the soil was compacted by the builders. More details of the study can be found here: https://www.gardenmyths.com/compost-tea-does-it-work/

The soil around the trees was treated with one of the following: water, ACT compost tea (aerated), a commercial bacterial product (CBP), compost mulch, or wood chip mulch.

Each tree received the same total amount of water, either as part of the treatment (water only, compost tea) or as a separate watering. Analysis of soil samples were done by independent labs. During the test period, the liquid additives were applied on a regular basis, and compost and wood chip thicknesses were renewed yearly.

The following chart shows the trees' growth after 5 years. Compost tea and the commercial microbes were no better than water alone.

How did the microbe population compare in each application? The following table shows the test results. CBP was a commercial biological product that was specifically sold and promoted for soil enhancement. This table shows the microbe populations in the material used to treat the trees. It does not take into account the actual amount added to each tree.

A big surprise for me was the microbe count on wood chips. They are not as high as ACT or compost, but they are significantly higher than I expected.

Keep in mind that people make compost tea to "concentrate"

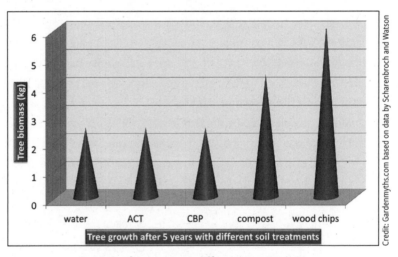

Growth of trees given different treatments.

Microbe Populations in Compost, Compost Tea, and Wood Chips.

	ACT	CBP	Compost	Wood Chips
Bacteria	2,688 mg/kg	200,000 mg/kg	4,698 mg/kg	1.000 mg/kg
Fungi	50 mg/kg	3,325 mg/kg	2,670 mg/kg	750 mg/kg
Flagellates	200/g	0	123,832/g	1,900/g
Amoebae	140/g	0	5,756/g	600/g
Ciliates	8/g	0	123/g	5/g
Nematodes	0.1/g	0	3/g	0.5/g

the microbes. They think they are increasing their numbers, but in fact compost has 50% more microbes than the tea.

The following table shows the amount of bacteria and fungi that were actually applied to the trees. The numbers are in kg/100 square meters/year. Other microbes such as ciliates are not expected to have a significant effect on soil or plants.

Amount of Microbes Used (kg/100 square meters/year).

	ACT	CBP	Compost	Wood Chips
Bacteria	0.57	0.42	8.22	4.05
Fungi	0.01	0.01	4.67	3.04

Soil quality was also measured using these parameters: density, moisture, organic matter, respiration, pH, nitrogen, phosphorus, and potassium. Density is a measure of the degree of compaction. A lower density indicates soil that is less compacted and of better quality.

Respiration is a measure of the amount of CO_2 produced. It is an indirect way of counting the microbes. A higher value indicates more microbes and a healthier soil.

ACT and CBP did improve the soil density but not more than compost or wood chips. As far as the other soil parameters go, ACT and CBP had limited effect. It is not surprising that ACT does not

Effect of Compost Tea, Compost, and Wood Chips on Soil

	ACT	CBP	Compost	Wood Chips
Density	Lower --	Lower -	Lower --	Lower --
Moisture	Same	Same	Higher +	Higher ++
Organic matter	Same	Same	Higher ++	Higher +
Respiration	Same	Same	Higher ++	Higher ++
pH	Same	Same	Higher ++	Higher +
Nitrogen	Same	Same	Higher ++	Higher +
Phosphorous	Same	Same	Higher +++	Higher +
Potassium	Same	Lower -	Higher +++	Higher ++

Credit: Gardenmyths.com based on data by Scharenbroch and Watson

add a lot of nutrients—it is mostly water, and CBP did not contain nutrients.

ACT does contain some microbes, but it is mostly water, and this is confirmed by the fact that it did not increase the level of organic matter.

What should be surprising to the proponents of compost tea is that respiration did not go up. This is a major claim for compost tea. It adds missing microbes to the soil and increases the level of both bacteria and fungi. The results of this study show that this is not true.

CBP is a product that claims to increase the number of microbes in soil which in turn improves the health of soil. The results clearly show that such products do not work.

Adding microbes to soil does not increase the number of microbes in the soil.

This is just one experiment, but there are others that show similar results. The science does not support most of the claims for compost tea. It does not increase the number of microbes in soil, and it does not improve soil health. Use the compost but skip making tea.

13

Selecting the Best Composting Method

Now that you have learned more about composting than you ever wanted to know, it's time to get composting. The next step is to select the best method. The best method is the one that suits your conditions most closely. Have a look at your garden and lifestyle and select the one that most closely meets your needs.

Don't be afraid to try several methods at the same time or change your methods over time. When I started gardening, I used mostly the trench method because I didn't have room for a compost pile. In later years, I used medium temperature piles. I never had enough material for full-sized piles. At my current garden, I have lots of space and lots of organic material so I built a two-bin hot composting system using skids. More recently I have stopped using it and have migrated to the cut and drop method because it is less work, and with a 6-acre garden, I have better ways to use my time. For kitchen scraps, I still use a small plastic dome composter.

Are you still undecided? Keep reading and I will help you make a decision.

Eco-enzyme is quite specialized and only suitable for smaller quantities of organic matter. You can do it on the side along with another method, but it won't handle the volume most people have.

Making compost tea is not a composting method, and it does not add much value to the garden.

That leaves traditional composting with all its variations, vermicomposting, and bokashi. The following table lists some key differences between these three methods. The focus is on home gardens and not farms. Go through the list and compare it to your goals and available environment. I think you will find a clear winner.

Feature	Traditional Composting	Vermicomposting	Bokashi
Need to manage pets	no	yes	no
Done outside	yes	normally done indoors	normally done indoors
Stinks	can stink a bit	no	has an odor that is not objectionable
Can be left alone on vacation	yes	for a limited time	yes
Recommended for meat and dairy	no	no	yes
Need to manage C/N ratio	yes	no	no
Works best for larger amounts of OM	yes	yes, but not indoors	no
Works best for kitchen scraps	no	yes	yes
Suitable for apartments	no	yes	yes
Can be done year-round	not in colder climates	yes, if indoors	yes
Cost	free or minimal cost	need to buy worms	need to buy on-going supplies
Kills weed seeds	yes, if hot	no	probably

Best Traditional Composting Method

If you eliminate worms and bokashi as possible methods, the next step is to figure out which outdoor composting method best matches your goals.

	Hot pile	Cold pile	Plastic composter	Tumbler	Cut & drop	Trench
Speed	fast	slow	slow	slow	doesn't matter	slow
Cost	low	low	higher	higher	low	low
Needs to be turned	yes	no	no	yes	no	no
C/N ratio is important	yes	no	yes	yes	no	no
Suitable for large quantities	yes	yes	no	no	yes with limits	yes with limits
Suitable for small quantities	no	yes	yes	yes	yes	yes
Easy to control moisture	yes	yes	no	no	doesn't matter	doesn't matter
Rodent proof	no	no	yes	yes	no	no
Easy	no	yes	yes	no	yes	yes

Environmental Concerns

Composting is considered to be a green activity and that it is good for the environment. As with all green activities, there is also a brown side to composting. All composting produces greenhouse gases, and the type of gas produced depends very much on the method you use.

When organic matter decomposes, microbes use the carbon source as energy and carry out a process called respiration. Chemically this is the opposite of photosynthesis. Sugars, the carbon source, are converted to water, energy, and CO_2. The CO_2 is released into the air and adds to global warming issues.

The process also requires nitrogen, and some of this is converted to nitrous oxide (N_2O), another greenhouse gas. Composting produces less nitrous oxide than CO_2, but N_2O has a larger effect on warming. One pound (one-half kilogram) of N_2O has the same warming effect as 300 pounds (135 kg) of CO_2.

Methane is also a greenhouse gas that is 30 times more effective at trapping heat than CO_2.

How significant is this problem? A study done by A. Boldrin, et al.,[1] looked at a popular backyard composting unit in Europe, the Humus Genplast, a green conical plastic composter. Kitchen waste is added from the top, and from time to time, it is stirred with a stick.

The average loss of carbon during composting was 70%, of which 73% was CO_2 and 2% was methane (CH4). The total loss of nitrogen (N) was 60%, of which 5% was nitrous oxide (N_2O). Loss of ammonia (NH3) was negligible.

Loss of carbon and nitrogen through the leachate was negligible.

As a side note, this very significant loss of carbon is the key reason why the compost pile shrinks so much during the process.

Composting clearly produces greenhouse gases, but what are the alternatives? The material can be sent to the municipal composting facility, but it will still produce the same amounts of CO_2. Depending on the process, it can also produce methane, not to mention the CO_2 produced to transport the input material and finished compost, as well as run the facility.

Some facilities in North America (Toronto, Ontario, being one) capture the methane and use it to run the operation, which is much better for the environment. Most facilities do not do this.

Why not just send it to the landfill? Organic matter in landfill goes through an anaerobic decomposition which produces something called landfill gas, which is approximately 50% CO_2 and 50% methane. In the United States, landfills are the third-largest source of human-related methane emissions, accounting for 15% of these emissions in 2019. That is equivalent to the greenhouse gas emissions from 22 million vehicles driven for one year. Sending organic material to landfill is our worst option.

Some composting methods are better for the environment than others, but there is little data on this yet. Consider two simple methods: cut and drop, and burying.

The cut and drop method uses very thin layers of material on top of the soil. This will produce very little methane because decomposition is very aerobic.

Burying the organic matter can be a problem. If it is buried too deep, it starts decomposing anaerobically, producing methane. The amount produced depends on the soil type. Coarse soils with good oxygen levels are less of a problem, but heavy clay soil with low oxygen produces more methane. Keeping the organic matter closer to the surface will help keep it aerobic.

In general, the more aerobic the process, the better it is for the environment.

Bokashi is anaerobic, but it is also not a composting process. Indications are that bokashi produces few greenhouse gases. The problem of course is that the ferment needs to compost at some point, and that is when the gases are released. A common way to do this is to bury the material, which may create higher levels of methane.

In the case of vermicomposting, I suspect worms do produce greenhouse gases during the digestion process, but most of the decomposition happens after castings are made. The gases produced will depend very much on how you handle the compost. Using it as a mulch should produce less methane than burying it.

In 2007 newspaper reports appeared with headlines something like this, "Worms Are Killing the Planet." This story was based on some new research which suggested that worms produce high levels of nitrous oxide, a greenhouse gas that is much worse than CO_2. This research has been reviewed and does not seem to have been done very well. The current consensus is that worms are not high producers of greenhouse gases.

A few studies have looked at the production of nitrous oxide and methane in home composting systems and have concluded that vermicomposting produces less than aerobic composting, which in turn produces less than anaerobic composting. This makes vermi-composting better for the environment.

When vermicomposting was compared with traditional hot composting in large agricultural settings, it was found to produce lower levels of methane and nitrous oxide, but CO_2 levels were about the same.

These types of studies generally exclude the production of CO_2 in calculating greenhouse gases because the amount of CO_2 produced is relative to the quantity of carbon in the starting waste material. Which means that the total CO_2 produced is about the same for most aerobic methods.

Is hot composting better than cold composting?

Standard hot aerobic composting produces CO_2 but very little methane or nitrous oxide. If the compost pile gets too wet, or if not enough air gets to the middle of the pile, it will start producing more methane and nitrous oxide. Nitrous oxide is also produced during the curing stage and after it is added to soil.

Cold composting is more anaerobic and is expected to produce more methane and nitrous oxide. Unfortunately, very few studies have looked at this, but hot composting is probably better for the environment.

Should You Compost?

Given the fact that composting creates greenhouse gases, should you compost? The answer is a clear yes because there are some positive aspects as well.

Taking plant material, composting it, and putting it in the ground does create some greenhouse gases, but it also adds carbon to the soil. This sequestered carbon reduces the CO_2 in the air. Remember that growing plants get their carbon by using CO_2 from the air. Composting returns some of that back to air, but the whole process is a net reduction of CO_2.

We all have kitchen and yard waste, and there are only two options. Keep it on the property or haul it away. Hauling it away still produces greenhouse gases as the material decomposes, but there is added pollution from transportation. Keeping it on your property is always a better option.

Creating your own compost also means that you need to use less purchased fertilizer. It does not matter if this fertilizer is synthetic or organic, both need to be produced and shipped around the country. Using your own compost reduces the greenhouse gases produced by manufacturing.

Mixing Composting Methods

Much of this book talks about different composting methods and helps you select the one that best meets your requirements, but this does not mean you have to select only one. Many of these methods can be combined, or run parallel to each other.

I mostly use the cut and drop method in the garden. I also have a plastic bin outside the garage for my kitchen scraps because, quite frankly, I don't want to see orange peels in the garden. I have also tried bokashi because it might be a good system to use in winter for kitchen scraps.

If you get a lot of fall leaves, consider making a cold compost pile, or use a wire bin for them.

A lot of smaller homes will have trouble collecting enough material for a hot compost pile, except maybe in fall or spring, when you clean up the garden. Use this method once a year and a different method the rest of the year.

Use the methods that work for you. Remember, any method is better than taking it to the curb.

14

Using Compost

The big day is finally here, and you have some finished compost. It's dark, moist, and crumbly—that black gold you've heard so much about. Now what? How do you use the stuff?

The first thing to realize is that not all composting methods produce a pile of compost. Bokashi produces ferment, and if you bury it, you will never have any compost. Keyhole gardening and the cut and drop method don't produce compost either. The organic matter just disappears into the soil. This is not a bad thing; it saves you from having to deal with it.

When Is Compost Finished?

Gardeners use different definitions to describe so-called finished compost, and I have discussed this along with some of the methods. The key point here is that the material gardeners consider to be finished is not really finished. It will continue to decompose for a long time. Compost is constantly changing, and it is not that critical for you to wait for a special end point. It can be used when you are ready to use it.

There is one potential problem you need to be aware of. Compost that is too fresh may harm plants because it releases too many nutrients at one time, or it has a very high carbon level that steals nitrogen from the soil. That is why most people let compost sit and age for a while. If in doubt, don't rush to use it.

Should You Sift Compost?

The material from compost piles is sifted to take out uncomposted material like small twigs, or large seeds like walnuts. These can be added to the next pile for further processing.

In vermicomposting, people sift the material to remove worms and cocoons so they can use them in a new bin.

Sifting is normally done using a wooden frame that has been fitted with a wire mesh screen. These can be simple devices that are held over a wheelbarrow, or you can make more complex systems that are suspended by a tripod.

Do you need to sift compost? It depends a bit on how you use it. If you use it as mulch, or dig it right into the garden, you don't have to sift it. The larger pieces will compost eventually, and any remaining worms will live or die. Sifting is not necessary in most cases.

If you plan to use the compost for potted plants, especially seedlings, it is better to sift it. If you are using vermicompost, most people don't appreciate worms hatching out of their houseplants, and a combination of sifting and drying will prevent this.

Mulching vs Burying

The obvious thing to do with compost is to dig it into the soil. That seems to be the quickest way to improve soil and benefit plants, but it may not be the best option.

Let's take a step back for a minute and have a look at the cause of poor quality soil. Soil left on its own does not degrade over time. The soil under trees in forests and soil under prairie grasses is all very good soil. Except for natural disasters, like landslides and flooding, the main reason for poor soil is agriculture. Using the land to grow crops slowly degrades the soil and removes nutrients.

We need farmers to produce our food, and they are very careful to take care of the soil as best they can, so don't blame farmers for this problem.

The main issue with agriculture has to do with tilling the soil. Each time soil is tilled, excess oxygen is added to it, and this oxygen activates microbes to speed up the decomposition of organic matter. This is exactly the same process that takes place in a hot compost pile where a lack of oxygen slows the process down and a high amount speeds it up. The same happens in soil.

In the long run, tilling reduces the level of organic matter, which in turn reduces the number of microbes the soil can support. Lower microbe levels lead to a destruction of soil structure.

Farmers have been aware of this for a long time, and they have moved to no-till or limited-till farming wherever they can. Working the soil less reduces the loss of organic matter and can even reverse the process.

It is ironic that gardeners are just now starting to understand the value of no-till gardening.

No-till gardening is great for soil, but if you practice this kind of gardening, how do you dig your compost into the soil? The short answer is that you don't. You mulch with it and leave it on top of the soil. All kinds of nature's critters, like worms and beetles, will help move it into the soil with much less damage to soil.

The study I presented in the chapter on compost tea used compost as mulch, and it significantly increased the soil organic matter. It will get into the soil on its own.

Mulch also keeps the soil cool, which roots prefer, and it helps maintain an even moisture level.

The one drawback with compost mulch is that it is a great seed bed. As weed seeds land on it, they have a perfect place to grow.

So what should you do with your compost, mulch with it or bury it?

If you are preparing a new bed and you plan to dig it up, or even rototill it, it's a good idea to apply a thick six-inch (150 mm) layer of compost before you dig up the garden. After this preparation stage, don't till the garden again. In future years, just apply compost as mulch.

Compost can also be spread on lawns at a one-quarter-inch (6 mm) depth. If you do this you should sift it first.

Beware of High Carbon Compost

The composting process requires a C/N ratio of about 30. If there is not enough nitrogen, composting is slow and/or incomplete. There is a potential problem if such high carbon compost is buried in soil. The soil microbes start to decompose it and, in the process, use up available nitrogen in the soil, taking it away from plants.

Compost that has a high amount of uncomposted leaves or wood chips falls into this high carbon category. It can be used as mulch, but should not be buried in soil.

Compost Myth: Wood Chips Rob Nitrogen

Do wood chips rob nitrogen from the soil? The answer depends on how you use them.

Wood chips have a very high C/N ratio, so in order for microbes to decompose them, they need to find nitrogen from another source and soil is their only option. If you bury wood chips, the microbes will take nitrogen from the soil and use it to decompose the wood. This in turn reduces the nitrogen available to plants and slows their growth. Eventually the material is decomposed, and the nitrogen is returned to the soil.

When wood chips are used as mulch, microbes still get nitrogen from the soil, but remember how small microbes are. They can only get nitrogen from the top fraction of an inch which is well above plant roots. Plants won't be affected by mulch.

Wood chips are a great mulch, and you don't have to worry about them robbing nitrogen from the soil.

Is Compost Safe for Vegetables?

Most compost is perfectly safe for use in a vegetable garden, but there are some special cases you should be aware of.

Any compost made from meat, pet waste, or even human waste may contain pathogens, unless it went through a truly hot composting phase. Personally, I think this risk is extremely low, but it is probably a good idea to keep such compost out of the vegetable garden. It can be used anywhere else. Human urine is not an issue.

Compost that is not fully finished, or one that has a high nitrogen level (e.g., animal manure), may be too "hot" to use, especially near seedlings in the vegetable garden. The word "hot" in this context is referring to the level of nutrients in the compost, not its temperature. High nutrient levels can burn tender roots, and that is more likely near seedlings.

Aging compost will reduce the nutrient level and make it safer. Using it as mulch also keeps it away from roots.

Amending Potting Soil

Potting soil is a soilless material that is usually based on peat moss or coir. It is used mostly for outdoor containers and indoor potted plants. This material has few if any nutrients in it.

Compost can be added to this material before potting plants to increase the nutrient level. It does not really increase the organic level since potting soil is almost 100% organic matter. The problem I have with doing this is that you have no idea about the level of nutrients you are adding because you don't know the nutrient level in the compost.

Compost releases nutrients slowly, and you may end up underfeeding plants. Or if the levels are high, you can harm your plants.

The reality is that synthetic fertilizers or organic fertilizers with a known NPK work better in containers because you can easily control the amount of fertilizer you add.

There is also a claim that the microbes in compost will "improve the soil." Keep in mind that potting soil is not soil, so there is no soil to improve. Adding microbes to potting soil is unlikely to have any benefits.

How Much Compost Should You Use?

Nobody can answer this question because they don't know the chemistry of your compost. It is best to start using small amounts and use more over time once you are confident you are not using too much.

As a starting point, add no more than 20% of fresh compost or 30% of well-aged compost to your potting mix. You can increase that to 60% over time. Some plants even grow well in 100% well-aged compost.

Compost can also be used as mulch on potted plants. Adding an inch at a time should be fine.

Can You Use Too Much Compost?

Compost is the best thing for your garden, so how can you have too much? The truth is that too much compost, especially manure compost, can be harmful to your soil and plants.

One of the benefits of compost is that it adds nutrients to soil. The amount of nutrients depends on how it is made and the input ingredients. Homemade compost, which is made mostly from plant material, has an NPK number of 3-0.5-1.5, and commercial manure compost has an NPK value of about 1-1-1. Compost based on manure tends to have a higher relative amount of phosphorus.

Plants absorb nutrients in the ratio of 3-1-2. They need more nitrogen than phosphate. If you use manure compost with a value of 1-1-1 and add enough to provide the needed nitrogen, you are adding three times too much phosphorus, which can't be used by the plant.

What happens to nitrogen and phosphorus in soil?

Nitrogen moves through soil fairly quickly and can be easily washed away by rain. Nitrogen can also be converted to N_2 and N_2O. Both are gases that escape into the air. Excess nitrogen that is not used by plants easily leaves the growing layer in the soil.

Phosphorus, on the other hand, moves very slowly through soil at a rate of an inch or two a year. It does not wash away easily, nor does it get converted to gasses. Excess phosphorus accumulates in the soil, and for the most part, it stays put.

Because of the different ways nitrogen and phosphorus move through soil, even plant-based compost will result in an accumulation of phosphorus. If this is done annually, there is a steady buildup of phosphorus levels in soil.

High phosphorus levels make it more difficult for plants to take up manganese and iron, resulting in deficiencies of these nutrients in the plant. This shows up as interveinal chlorosis of the leaves. Some people try to solve this problem by adding more iron to the soil, but if the problem is caused by too much phosphorus, the last thing the soil needs is more iron.

High phosphorus levels are also toxic to mycorrhizal fungi which are very important to landscape plants. They provide phosphorus and water, as well as other nutrients, to the plant. Without mycorrhizal fungi, plants need to expend more energy making larger root systems. Less energy is then available for growing, flowering, and fruiting.

If phosphorus levels build up even more, it becomes toxic to plants.

Adding a six-inch (150 mm) layer of compost to poor soil at the time you make the initial garden bed will not be too much. An annual one-to-two-inch mulch layer is also fine.

Raised beds have become very popular, and some people are filling them with compost, or they are using soil plus compost where the compost portion is in the order of 50%. This might work, but we are also seeing poor growth and stunted plants in such systems. When the soil is tested, it invariably shows toxic levels of phosphate.

What can you do with soil that has toxic levels of nutrients? You can water a lot. That washes some nutrients through the bottom of the bed, but it does not wash phosphate out very well. You can also grow plants and remove them from the garden. This slowly reduces nutrient levels. Probably your best option is to take the bed apart, remove most of the soil, and replace it.

Storing Compost

I have discussed storing vermicompost in its own chapter. It can be stored and used over time. Storing other forms of compost is less advisable. It is always best to use the compost as soon as it is ready. Past a certain point, compost does not get better with age.

If you want to store compost, place it where it is not exposed to rain and let it dry out slowly. This will slow down decomposition and preserve nutrients. Garbage cans work well for this.

Compost Myth: Stored Compost Can Go Bad

Can compost go bad, especially when it's stored? To answer this, you first have to define what you mean by "bad." Since the purpose of compost is to improve soil, bad compost would either harm soil or lose the ability to improve soil.

A comment I found online said, "Excess water or moisture that collects deep in the pile can cause the compost to rot, so that it can't be used on your soil."

But do we not want it to rot, since composting is in fact a rotting process?

When compost is stored, it continues to decompose, so it is slowly losing nutrients and carbon. Over time you can see the amount slowly decrease. This does not make it bad—it simply means it is more decomposed.

Compost that is stored without enough air exchange can become anaerobic, and different microbes might start to grow, causing it to stink. If you add this to soil, it will have access to oxygen and aerobic conditions will return. This is especially true if you use it as mulch. So stinky compost is not bad.

The chance of anaerobic conditions is further reduced if the compost is dried out before storing.

Over time compost does lose nutrients and carbon, and that is why it is a good idea to use it as soon as it is finished. However, compost can't go bad.

When Is the Best Time to Use Compost?

This one is simple, as soon as it's ready.

Compost that is too fresh might release too many nutrients, but this is rarely a problem if it is used as mulch or placed away from plant roots. You can always check it with a Seedling Test.

Is it better to apply in spring or fall?

Think of compost as a slow-release fertilizer that is continually releasing nutrients. The best time to add it is when plants are growing, so spring is a great time because it will feed plants all summer. If you apply it in fall or winter, the temperatures are dropping and microbes become less active, which means nutrients are released more slowly. For this reason, an application in late fall is about the same as an early spring application.

In warm climates, compost can be used any time that plants are growing.

Keep in mind that roots of permanent plantings, like perennials, trees, and shrubs, do most of their growing in cool weather, and they also need nutrients.

I would not be too concerned about when you apply compost. Use it as soon as you have it.

Can Compost Make You Sick?

Compost sounds like such a great thing, but it can make you sick. The chance of this happening is very small, but it is possible. People have died from it.

The first thing to be concerned about is the dust. Small dust particles can coat your lungs and cause breathing issues. This is a bigger concern with dry compost, so it is a good idea to wet it down before you use it. It is also a good idea to wear a respirator.

There are also a number of diseases that you can catch from compost. The best way to protect yourself is with masks and gloves. It is important to put these diseases into perspective—they are all extremely rare. I do get a tetanus shot, and I do wear gloves to prevent infections on my hands. I don't wear a respirator.

Tetanus

Tetanus is an infection caused by a bacterium called *Clostridium tetani*, which is present everywhere in the environment, including soil, manure, and compost. The spores can get into the body through broken skin while working in the garden.

The illness manifests itself in three to twenty-one days after infection with an average of ten days. This results in sustained muscle contractions of the jaw, hence the common name lockjaw. Spasms of the jaw or facial muscles may follow, spreading to the hands, arms, legs, and back, and blocking the ability to breathe. Tetanus is not contagious from person to person.

The best way to prevent this disease is to be immunized against it.

A rusty nail does not really cause tetanus, but the wound left by the nail can become infected and rusty nails are usually dirty.

Legionnaires' Disease

Legionnaires' disease is the result of infection from *Legionella* bacteria which cause pneumonia-like symptoms. There are some 42 species of this bacteria, and 18 have been linked to pneumonia-like infections in humans.

The EPA reports that, "Legionella are ubiquitous in natural aquatic environments, capable of existing in waters with varied temperatures, pH levels, and nutrient and oxygen contents. They can be found in groundwater as well as fresh and marine surface waters."

Most cases of pneumonia are not tested for Legionnaires' disease, so we don't know the actual number of cases. Some estimates suggest that only 10% are diagnosed correctly.

There have been reports in New Zealand, Europe, and North America that bagged potting soil caused Legionnaires' disease. As people move away from peat-based potting mixes and use wood-based ones instead, there has been an increase in cases.

Legionella bacteria are certainly found in soil and potting mix. It is quite possible that it also exists in compost, although I have not seen a direct connection yet.

The disease seems to affect certain individuals, and many others seem to be immune. The factors that contribute to this are mostly unknown. Smokers, heavy drinkers, and people with immune diseases seem to be more susceptible, but the data about this is limited, mostly because very few cases have been studied.

Wood-based potting mix may increase the risk. Compost is more likely to spread the disease if it has high levels of moisture and nutrients, the perfect place for bacteria to grow. However, the risk of this disease is still very low.

Farmer's Lung

Farmer's lung is an allergic reaction to some types of mold in certain plants, such as hay, corn, grass for animal feed, grain, and tobacco. It is caused by breathing in dust which causes inflammation or swelling on the lungs.

Symptoms usually show four to eight hours after exposure and can include dry cough, chills, rapid breathing, rapid heart rate, shortness of breath, and a general feeling that you're sick.

Histoplasmosis

Histoplasmosis is a disease caused by a fungus called *Histoplasma capsulatum*. It usually affects the lungs, but it can also infect eyes, skin, and adrenal glands. Symptoms are similar to Farmer's lung. They are normally mild; however, histoplasmosis can be severe and produce an illness similar to tuberculosis.

The organism thrives at moderate temperatures in rich moist soils, especially ones containing bird droppings.

Epilogue: Beware of Garden Myths

In an effort to research the material for this book, I joined some Facebook groups dedicated to various types of composting. I understood the more technical aspects of the subject but wanted to better understand what actual users had to say.

It was quite an eye-opener. These are almost cult-like groups with very specific rules about what you can or cannot do to make compost. The problem is that many of their ideas are wrong and not science based.

One comment from a couple of days ago claimed it was a good idea to take compost from a hot composting pile and feed it to worms in a vermicompost bin. So I asked why?

When I see claims that don't make sense to me, I like to ask the author to explain the reason behind them. It might uncover a new way of thinking, but usually it shows that the author has no basis for the claim.

In this case I was told that the worms concentrate the minerals in the compost, making it even more nutritious than just plain compost. Worms are given all kinds of superpowers by their owners.

Worms can't create more mineral nutrients. Whatever was in the worm bin and in the added compost is the total amount in the final vermicompost. That is basic physics. If anything, the process will probably decrease the amount of nitrogen, and maybe even sulfur, because they can become volatile.

Feeding compost to worms makes no sense at all.

As you continue in your journey of learning be very aware of myths. You will find them in every hobby and topic. Always ask why and be skeptical about the response.

I formed a Facebook group called Garden Fundamentals where we focus on science-based information. You are free to join and be part of our myth-busting group.

A few days ago in that group, someone asked about adding compost to a newly planted tree. This has been the accepted practice for many years, but is no longer considered the correct way to plant trees. Just replace the original soil and add nothing, not even fertilizer.

Someone posted that they have added compost to all of their newly planted trees and none died. This was their evidence that "it worked."

This illustrates a fundamental flaw in thinking. Just because compost did not kill the trees does not mean it was the best option. The trees might have done even better without the compost.

This is a good example of the need for controls. Gardeners tend to have one of everything. They plant one apple tree. If they do plant a second apple tree, it will probably be a different kind. There are no controls to test anything.

Research will use many of the same tree and split them into two groups. One set is planted with compost, and the other is planted without it. The trees are then monitored and measured for growth over several years. The trees without the compost are the controls. Comparing the control to the test subjects allows you to evaluate the benefits of adding compost.

Gardeners rarely use controls. It is more expensive and requires more space, so it's no wonder they don't use them, but without the controls, they can't reach logical conclusions. Unfortunately, they go with their gut feelings and do announce conclusions.

I like to illustrate the point with this story. My viburnum was hit badly with viburnum beetles, which decimated the leaves. This happened every year for three years. In spring of the fourth year, I took a lawn chair and a beer and sat next to the shrub while I enjoyed the beer. I had no viburnum beetle that year. Drinking beer prevented the beetle from laying eggs.

That is a pretty silly story, but every part of it, except the beer drinking, is true. I did nothing to solve the problem. If I had sprayed the bush with beer and had no beetle damage, I could have gotten online and told the world about how great my natural pesticide is. A new myth would be born.

Why does this happen? Because gardeners do not understand the need for controls. You could use two bushes and spray one and not the other. Or you could spray half the bush and not the other half. If you then see a significant number of beetles on one side, you can start to make a claim, but even more testing would be needed to validate it.

Many of the myths I see are a direct result of people reaching conclusions without any valid information to support their claim. You need controls, and you need to measure the outcome.

Take the time to understand the scientific method, and it will make you a better gardener, but even more important, it will allow you to make better life decisions. Science is important in all aspects of life.

Endnotes

Chapter 2: The Role of Compost in Soil

1. Bryant Scharenbroch, G.W. Watson, "Wood Chips and Compost Improve Soil Quality and Increase Growth of Acer rubrum and Betula nigra in Compacted Urban Soil," *Arboriculture & Urban Forestry* 40(6):319–331, November 2014.

Chapter 4: Compostable Material

1. Environmental Protection Agency, "Registration Review of Pyridine and Pyrimidine Herbicides," epa.gov/ingredients-used-pesticide-products/-registration-review-pyridine-and-pyrimidine-herbicides.
2. European Community Glyphosate Renewal Group, "Procedure and Outcome of the Draft Renewal Assessment Report on Glyphosate," June 2021.
3. Chunyang Liao and Kurunthachalam Kannan, "Widespread Occurrence of Bisphenol A in Paper and Paper Products: Implications for Human Exposure," *Environmental Science and Technology*, 45(21): 9372–9, 2011.
4. Government of Canada, "Final Human Health State of the Science Report on Lead," 2013, canada.ca/en/health-canada/services/environmental-workplace-health/reports-publications/environmental-contaminants/-final-human-health-state-science-report-lead.html
5. Sally L. Brown, Rufus L. Chaney, and Ganga M. Hettiarachchi, "Lead in Urban Soils: A Real or Perceived Concern for Urban Agriculture?," *Journal of Environmental Quality*, Volume 45, Issue 1, 2016.

Chapter 6: Piles, Bins and Tumblers

1. Khurram Ahahzad, et al., "Carbon Dioxide and Oxygen Exchange at the Soil-atmosphere Boundary as Affected by Various Mulch Materials," *Soil and Tillage Research* 194:104335, 2019.

Chapter 8: Vermicomposting

1. John Allen, "Vermicomposting, New Mexico State University, Guide H-164," aces.nmsu.edu/pubs/_h/H164/
2. Su Lin Lim, Ta Yeong Wu, Pei Nie Lim and Katrina Pui Yee Shak, "The Use of Vermicompost in Organic Farming: Overview, Effects on Soil and Economics," Wiley Online Library: 26 August 2014.

Chapter 9: Bokashi Composting

1. Håkan Asp and Helena Karlén, *Evaluation of Bokashi Fermentation Leachate as a Biofertilizer in Urban Horticulture*, Faculty of Landscape Architecture, Horticulture and Crop Production Science, stud.epsilon .slu.se/7353/11/lind_0_140929.pdf

Chapter 11: Buying Compost

1. Carlos Edo, Francisco Fernández-Piñas, Roberto Rosal, "Micro-plastics Identification and Quantification in the Composted Organic Fraction of Municipal Solid Waste," *Science of the Total Environment*, Volume 813, 20 March 2022.

2. EPA, Report on Priority Microplastics Research Needs: Update to the 2017 Microplastics Expert Workshop, 2021, epa.gov/trash-free-waters /science-case-studies

3. Jonathan D. Judy, et al., "Microplastics in Municipal Mixed-waste Organic Outputs Induce Minimal Short to Long-term Toxicity in Key Terrestrial Biota," 2019, (Pt A):522-531.

4. Anderson Abel de Souza Machado, Chung W. Lau, Werner Kloas, Joana Bergmann, Julien B. Bachelier, Erik Faltin, Roland Becker, Anna S. Görlich, and Matthias C. Rillig, "Microplastics Can Change Soil Properties and Affect Plant Performance," *Environ. Sci. Technol.*, 53, 10:6044–6052, 2019

5. Environmental Protection Agency, "Registration Review of Pyridine and Pyrimidine Herbicides," epa.gov/ingredients-used-pesticide -products/-registration-review-pyridine-and-pyrimidine-herbicides

6. Jill Crossman, Rachel R.Hurley, Martyn Futter, and LucaNizzetto, "Transfer and Transport of Microplastics from Biosolids to Agricul-tural Soils and the Wider Environment," *Science of the Total Environ-ment*, Volume 724, 1, July 2020.

Chapter 12: Compost Tea

1. Steven J. Scheuerell and Walter F Mahaffee, "Variability Associated with Suppression of Gray Mold (Botrytis cinerea) on Geranium by Foliar Applications of Nonaerated and Aerated Compost Teas," *Plant Disease*, 90(9):1201-1208, Sep 2006.

2. Vern Grubinger, "Compost Tea to Suppress Plant Disease," Univer-sity of Vermont, 2005, uvm.edu/vtvegandberry/factsheets/compost tea.html

3. WSU—compost tea, https://puyallup.wsu.edu/lcs/reference-compost -tea/

4. Bryant C. Scharenbroch and Gary W. Watson. "Wood Chips and Compost Improve Soil Quality and Increase Growth of Acer Rubrum

and Betula Nigra in Compacted Urban Soil," *Arboriculture & Urban Forestry*, 40(6): 319–331, 2014.

Chapter 13: Selecting the Best Composting Method

1. A. Boldrin, et al., "Mass Balances and Life Cycle Inventory of Home Composting of Organic Waste," *Waste Management*, 31(9-10):1934–42, June 2011, researchgate.net/publication/51206438_Mass_bal ances _and_life_cycle_inventory_of_home_composting_of_organic_waste)

Index

About the Author

ROBERT PAVLIS, a Master Gardener with over 45 years of gardening experience, is owner and developer of Aspen Grove Gardens, a six-acre botanical garden featuring 3,000 varieties of plants. A popular and well-respected speaker and teacher, Robert has published articles in *Mother Earth News*, *Ontario Gardening* magazine, a monthly "Plant of the Month" column for the Ontario Rock Garden Society website, and local newspapers. He is also the author of two widely read blogs: GardenMyths.com, which explodes common gardening myths; and GardenFundamentals.com, which provides gardening and garden design information. Robert also has a gardening YouTube channel called Garden Fundamentals.

Connect with Robert Pavlis

You can connect with me through social media by leaving comments at one of the following:

- http://www.gardenmyths.com
- http://www.gardenfundamentals.com
- http://www.youtube.com/Gardenfundamentals1

The best way to reach me directly is through my Facebook Group: https://www.facebook.com/groups/GardenFundamentals, where I answer questions on a daily basis.

Also by the Author

Plant Science for Gardeners

Plant Science for Gardeners empowers growers to analyze common problems, find solutions, and make better decisions in the garden for optimal plant health and productivity.

By understanding the basic biology of how plants grow, you can become a *thinking* gardener with the confidence to problem solve for optimized plant health and productivity. Learn the science and ditch the rules! Coverage includes:

- The biology of roots, stems, leaves, and flowers
- Understanding how plants function as whole organisms
- The role of nutrients and inputs
- Vegetables, flowers, grasses, and trees and shrubs
- Propagation and genetics
- Sidebars that explode common gardening myths
- Tips for evaluating plant problems and finding solutions.

Whether you're a home gardener, micro-farmer, market gardener, or homesteader, this entertaining and accessible guide shortens the learning curve and gives you the knowledge to succeed no matter where you live.

Soil Science
for Gardeners

Robert Pavlis, a gardener for over four decades, debunks common soil myths, explores the rhizosphere, and provides a personalized soil fertility improvement program in this three-part popular science guidebook. Coverage includes:

- Soil biology and chemistry and how plants and soil interact
- Common soil health problems, including analyzing soil's fertility and plant nutrients
- The creation of a personalized plan for improving your soil fertility, including setting priorities and goals in a cost-effective, realistic time frame.
- Creating the optimal conditions for nature to do the heavy lifting of building soil fertility.

Written for the home gardener, market gardener, and micro-farmer, *Soil Science for Gardeners* is packed with information to help you grow thriving plants.

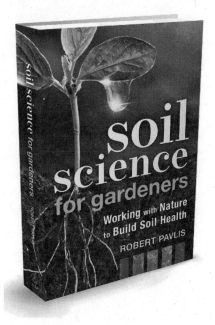

Garden Myths

If you enjoyed this book, you may also like Robert's other books, *Garden Myths Book 1 and Garden Myths Book 2*. Each one examines over 120 horticultural urban legends.

Turning wisdom on its head, Robert Pavlis dives deep into traditional garden advice and debunks the myths and misconceptions that abound. He asks critical questions and uses science-based information to understand plants and their environment. Armed with the truth, Robert then turns this knowledge into easy-to-follow advice. Details about the book can be found at http://www.gardenmyths.com/garden-myths-book-1/ .They are available from Amazon, and other online outlets.

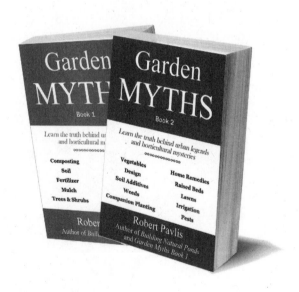

Building Natural Ponds

Building Natural Ponds is the first step-by-step guide to designing and building natural ponds that use no pumps, filters, chemicals, or electricity and mimic native ponds in both aesthetics and functionality. Highly illustrated with how-to drawings and photographs.

For more information and ordering details, visit:
www.BuildingNaturalPonds.com

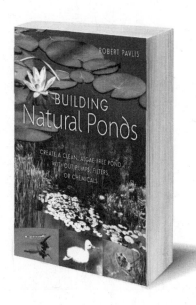